Active Learning

Synthesis Lectures on Artificial Intelligence and Machine Learning

Editor
Ronald J. Brachman, *Yahoo! Research*
William W. Cohen, *Carnegie Mellon University*
Thomas Dietterich, *Oregon State University*

Active Learning

Burr Settles

ISBN: 978-3-031-00432-2   paperback
ISBN: 978-3-031-01560-1   ebook

DOI 10.1007/978-3-031-01560-1

A Publication in the Springer series
SYNTHESIS LECTURES ON ARTIFICIAL INTELLIGENCE AND MACHINE LEARNING

Lecture #18
Series Editors: Ronald J. Brachman, *Yahoo Research*
                William W. Cohen, *Carnegie Mellon University*
                Thomas Dietterich, *Oregon State University*
Series ISSN
Synthesis Lectures on Artificial Intelligence and Machine Learning
Print 1939-4608   Electronic 1939-4616

Active Learning

Burr Settles
Carnegie Mellon University

SYNTHESIS LECTURES ON ARTIFICIAL INTELLIGENCE AND MACHINE LEARNING #18

ABSTRACT

The key idea behind active learning is that a machine learning algorithm can perform better with less training if it is allowed to *choose* the data from which it learns. An active learner may pose "queries," usually in the form of unlabeled data instances to be labeled by an "oracle" (e.g., a human annotator) that already understands the nature of the problem. This sort of approach is well-motivated in many modern machine learning and data mining applications, where unlabeled data may be abundant or easy to come by, but training labels are difficult, time-consuming, or expensive to obtain.

This book is a general introduction to active learning. It outlines several scenarios in which queries might be formulated, and details many query selection algorithms which have been organized into four broad categories, or "query selection frameworks." We also touch on some of the theoretical foundations of active learning, and conclude with an overview of the strengths and weaknesses of these approaches in practice, including a summary of ongoing work to address these open challenges and opportunities.

KEYWORDS

active learning, expected error reduction, hierarchical sampling, optimal experimental design, query by committee, query by disagreement, query learning, uncertainty sampling, variance reduction

Dedicated to my family and friends,
who keep me asking questions.

Contents

Preface

Machine learning is the study of computer systems that improve through experience. Active learning is the study of machine learning systems that improve by asking questions. So why ask questions? (Good question.) The key hypothesis is that if the learner is allowed to choose the data from which it learns — to be active, curious, or exploratory, if you will — it can perform better with less training. Consider that in order for most supervised machine learning systems to perform well they must often be trained on many hundreds or thousands of labeled data instances. Sometimes these labels come at little or no cost, but for many real-world applications, labeling is a difficult, time-consuming, or expensive process. Fortunately in today's data-drenched society, unlabeled data are often abundant (or at least easier to acquire). This suggests that much can be gained by using active learning systems to ask effective questions, exploring the most informative nooks and crannies of a vast data landscape (rather than randomly and expensively sampling data from the domain).

This book was written with students, researchers, and other practitioners of machine learning in mind. It will be most useful to those who are already familiar with the basics of machine learning and are looking for a thorough but gentle introduction to active learning techniques. We will assume a basic familiarity with probability and statistics, some linear algebra, and common supervised learning algorithms. An introductory text in artificial intelligence (Russell and Norvig, 2003) or machine learning (Bishop, 2006; Duda et al., 2001; Mitchell, 1997) is probably sufficient background. Ardent students of computational learning theory might find themselves annoyed at the lack of rigorous mathematical analysis in this book. This is partially because, until very recently, there has been little interaction between the sub-communities of theory and practice within active learning. While some discussion of underlying theory can be found in Chapter 6, most of this volume is focused on algorithms at a qualitative level, motivated by issues of practice.

The presentation includes a mix of contrived, illustrative examples as well as benchmark-style evaluations that compare and contrast various algorithms on real data sets. However, I caution the reader not to take any of these results at face value, as there are many factors at play when choosing an active learning approach. It is my hope that this book does a good job of pointing out all the subtleties at play, and helps the reader gain some intuition about which approaches are most appropriate for the task at hand.

This active learning book is the synthesis of a previous literature survey (Settles, 2009) with material from other lectures and talks I have given on the subject. It is meant to be used as an introduction and reference for researchers, or as a supplementary text for courses in machine learning — supporting a week or two of lectures — rather than as a textbook for a complete full-term course on active learning. (Despite two decades of research, I am not sure that there is enough breadth or depth of understanding to warrant a full-semester course dedicated to active learning. At least not yet!) Here is a road map:

- Chapter 1 introduces the basic idea of, and motivations for, active learning.

- Chapters 2–5 focus on different "query frameworks," or families of active learning heuristics. These include several algorithms each.

- Chapter 6 covers some of the theoretical foundations of active learning.

- Chapter 7 summarizes the various pros and cons of algorithms covered in this book. It outlines several important considerations for active learning in practice, and discusses recent work aimed at addressing these practical issues.

I have attempted to wade through the thicket of papers and distill active learning approaches into core conceptual categories, characterizing their strengths and weaknesses in both theory and practice. I hope you enjoy it and find it useful in your work.

Supplementary materials, as well as a mailing list, links to video lectures, software implementations, and other resources for active learning can be found online at http://active-learning.net.

Burr Settles
May 2012

Acknowledgments

This book has a roundabout history, and there are a lot of people to thank along the way. It grew out of an informal literature survey I wrote on active learning (Settles, 2009) which in turn began as a chapter in my PhD thesis. During that phase of my career I am indebted to my committee, Mark Craven, Jude Shavlik, Xiaojin "Jerry" Zhu, David Page, and Lewis Friedland, who encouraged me to expand on my review and make it publicly available. There has been a lot of work in active learning over the past two decades, from simple heuristics to complex and crazy ideas coming from a variety of subfields in AI and statistics. The survey was my attempt to curate, organize, and make sense of it for myself; to help me understand how my work fit into the overall landscape.

Thanks to John Langford, who mentioned the survey on his popular machine learning blog[1]. As a result, many other people found it and found it helpful as well. Several people encouraged me to write this book. To that end, Jude Shavlik and Edith Law (independently) introduced me to Michael Morgan. Thanks to Michael, William Cohen, Tom Dietterich, and others at Morgan & Claypool for doing their best to keep things on schedule, and for patiently encouraging me through the process of expanding what was a literature review into more of a tutorial or textbook. Thanks also to Tom Mitchell for his support and helpful advice on how to organize and write a book.

Special thanks to Steve Hanneke and Sanjoy Dasgupta for the detailed feedback on both the original survey and the expanded manuscript. Chapter 6 is particularly indebted to their comments as well as their research. I also found Dasgupta's review of active learning from a theoretical perspective (Dasgupta, 2010) quite helpful. The insights and organization of ideas presented here are not wholly my own, but draw on conversations I have had with numerous people. In addition to the names mentioned above, I would like to thank Josh Attenberg, Jason Baldridge, Carla Brodley, Aron Culotta, Pinar Donmez, Miroslav Dudík, Gregory Druck, Jacob Eisenstein, Russ Greiner, Carlos Guestrin, Robbie Haertel, Ashish Kapoor, Percy Liang, Andrew McCallum, Prem Melville, Clare Monteleoni, Ray Mooney, Foster Provost, Soumya Ray, Eric Ringger, Teddy Seidenfeld, Kevin Small, Partha Talukdar, Katrin Tomanek, Byron Wallace, and other colleagues for turning me on to papers, ideas, and perspectives that I might have otherwise overlooked. I am sure there are other names I have forgotten to list here, but know that I appreciate all the ongoing discussions on active learning (and machine learning in general), both online and in person. Thanks also to Daniel Hsu, Eric Baum, Nicholas Roy, and their coauthors (some listed above) for kindly allowing me to reuse figures from their publications.

I would like to thank my parents for getting me started, and my wife Natalie for keeping me going. She remained amazingly supportive during my long hours of writing (and re-writing).

[1] http://hunch.net

Whenever I was stumped or frustrated, she was quick to offer a fresh perspective: "Look at you, you're writing a book!" Lo and behold, I have written a book. I hope you enjoy the book.

While writing this book, I was supported by the Defense Advanced Research Projects Agency (under contracts FA8750-08-1-0009 and AF8750-09-C-0179), the National Science Foundation (under grant IIS-0968487), and Google. The text also includes material written while I was supported by a grant from National Human Genome Research Institute (HGRI). Any opinions, findings and conclusions, or recommendations expressed in this material are mine and do not necessarily reflect those of the sponsors.

Burr Settles
May 2012

CHAPTER 1

Automating Inquiry

"Computers are useless. They can only give you answers."

— *Pablo Picasso (attributed)*

1.1   A THOUGHT EXPERIMENT

Imagine that you are the leader of a colonial expedition from Earth to an extrasolar planet. Luckily, this planet is habitable and has a fair amount of vegetation suitable for feeding your group. Importantly, the most abundant source of food comes from a plant whose fruits are sometimes smooth and round, but sometimes bumpy and irregular. See Figure 1.1 for some examples.

Figure 1.1: Several alien fruits, which vary in shape from round to irregular.

Almost immediately, physicians on the expedition notice that colonists who eat the smooth fruits find them delicious, while those who eat the irregular fruits tend to fall ill. Naturally, you want to be able to distinguish between which fruits are safe to eat and which ones are harmful. If you can accurately "classify" these fruits, it will be much easier to keep the colony healthy and well-fed while finding important alternative uses for the noxious fruits (such as processing them into dyes and fuels). The physicians assure you that the shape of a fruit is the only feature that seems related to its safety. The problem, though, is that a wide variety of fruit shapes from these plants exist: almost a continuous range from round to irregular. Since the colony has essential uses for both safe and noxious fruits, you want to be able to classify them as accurately as possible.

Supposing we have a way to quantify the "irregularity" of a fruit's shape, we can formalize this classification task using a simple function. Let $\mathcal{X}$ be a range of data instances, e.g., fruits that have been harvested, where each $x \in \mathbb{R}$ represents the irregularity measure of a particular fruit. Each data instance has a hidden label y, which in this example comes from the finite set $\mathcal{Y} = \{\text{safe, noxious}\}$.

Our classifier, then, is a function mapping $h : \mathcal{X} \to \mathcal{Y}$, parameterized by a threshold θ:

$$h(x; \theta) = \begin{cases} \oplus \text{ safe} & \text{if } x < \theta, \text{ and} \\ \ominus \text{ noxious} & \text{otherwise.} \end{cases}$$

The remaining challenge is to figure out what this threshold really is.

One way to pick the threshold is by *supervised learning* techniques, where we estimate θ from a set of *labeled* data instances $\mathcal{L} = \{\langle x, y \rangle^{(l)}\}_{l=1}^{L}$. Here, $\langle x, y \rangle$ denotes a fruit x which has been tested and assigned a label y, e.g., by having someone eat it and then observing whether or not he or she becomes ill. A simple learning algorithm in this case is to harvest a large number of fruits, arrange them in order from round to irregular in shape, and have them all tested. The point at which the fruits switch from being safe to noxious is the best threshold θ^* (assuming for the moment that everyone responds to a fruit in the same way). Figure 1.2 illustrates this method for a handful of labeled instances.

Figure 1.2: Supervised learning for the alien fruits example. Given a set of $\langle x, y \rangle$ instance-label pairs, we want to choose the threshold θ^* that classifies them most accurately.

According to the *PAC learning*[1] framework (Valiant, 1984), if the underlying data distribution can be perfectly classified by some hypothesis h in the hypothesis class $\mathcal{H}$ (in this case, the set of all values for θ), then it is enough to test $O(1/\epsilon)$ randomly selected instances, where ϵ is the maximum desired error rate. This means that if we want to achieve 99% accuracy in our "fruit safety" classifier, we might need to harvest and test on the order of hundreds of fruits using this approach. That is a lot of fruits to test, though, and many of your family and friends might get sick in the process!

Can we do better? While we certainly want a threshold that classifies these fruits as accurately as possible, we also want to discover this threshold as *quickly* and as *economically* as possible, by requiring fewer tests while achieving the same results (ideally). Note some key properties of this problem:

- the fruits themselves are plentiful, easily harvested, and easily measured;

- testing for a fruit's safety is what mainly incurs a cost: the fruit must be consumed, and the person who eats it risks falling ill.

---

[1]More detail on PAC learning and active learning will be discussed in Chapter 6.

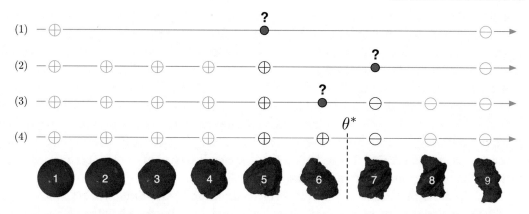

Figure 1.3: A binary search through the set of ordered, untested alien fruits. By only testing this subset of fruits, we can exponentially reduce costs while achieving the same result. The labels shown in light blue can be inferred, and therefore do not need to be tested.

As it turns out, we *can* do better by performing a directed search through the set of fruit shapes. Taking the two observations above into account, we can augment our supervised learning strategy in the following way. First, let us gather an arbitrary number of *unlabeled* instances $\mathcal{U} = \{x^{(u)}\}_{u=1}^{U}$ for free (or very inexpensively) by simply harvesting a lot of fruits without testing them. Next, we can arrange these fruits in a sequence from round to irregular. Our goal is similar to the previous one: to discover where in this ordered series of fruits the transition from **safe** to **noxious** occurs, but by conducting as few tests as possible this time. If we execute a simple binary search through this ordered set of fruits, we only need to test $\lceil \log_2 U \rceil$ items, instead of testing them all as we did before.

The process is illustrated in Figure 1.3. Given a set of unlabeled instances, this sequential algorithm would first test $x = 5$, then $x = 7$, and finally $x = 6$ before arriving at the exact same parameter value for θ^*, alleviating the need to test the other six fruits (two of which happen to be harmful). This algorithmic speedup means that a classifier with error ϵ or less can be found with a mere $O(\log_2 1/\epsilon)$ tests, since all the other labels can be inferred. To put this in perspective, assume that 100 fruits were needed to obtain 99% accuracy in the earlier supervised setting. Then we would still need to harvest 100 fruits for our new binary search algorithm, but we would only need to test 6 or 7 of them to get the same guarantee. This is an *exponential* reduction in the number of tests necessary, which drastically reduces the number of people whose health is at risk, and helps the colony to make use of both safe and noxious fruits as quickly as possible.

1.2   ACTIVE LEARNING

The alien fruits example is a simple illustration of *active learning*. Active learning is a subfield of artificial intelligence and machine learning: the study of computer systems that improve with experience and training. Traditional "passive" learning systems induce a hypothesis to explain whatever training data happens to be available (e.g., a collection of labeled instances). By contrast, the hallmark of an active learning system is that it eagerly develops and tests new hypotheses as part of a continuing, interactive learning process. Another way to think about it is that active learners develop a "line of inquiry," much in the way a scientist would design a series of experiments to help him or her draw conclusions as efficiently as possible. For example, the binary search algorithm in Figure 1.3 selects the next fruit to test — or *query* — after it has obtained an answer for the previous query from some *oracle* (or labeling source). For this reason, active learning is sometimes called "query learning" in the computational learning theory literature, and is closely related to work in "optimal experimental design" or "sequential design" in statistics.

Of course, the alien fruits example is a bit contrived and overly simplistic. Yet it illustrates some of the basic properties of many real-world problems, and shows how much can be gained from having the learner ask questions or be more involved in its own training process. In today's information-drenched society, unlabeled data are often abundant, but the labels required to do supervised learning from these instances are difficult, time-consuming, or expensive to obtain. Consider a few examples:

- *Classification and filtering.* Learning to classify documents (e.g., articles or web pages) or any other kind of media (e.g., image, audio, and video files) usually requires people to annotate each item with particular labels, such as relevant or not-relevant. Unlabeled instances abound from electronic resources like the Internet, but annotating thousands of these instances can be tedious and even redundant.

- *Speech recognition.* Accurate labeling of speech utterances is extremely time consuming and requires trained linguists. While unannotated audio data is easily obtained from recording devices, Zhu and Goldberg (2009) have reported that annotation at the word level can take ten times longer than the actual audio (e.g., one minute of speech takes ten minutes to label), and annotating phonemes can take 400 times as long (e.g., nearly seven hours). The problem is compounded for rare languages or dialects.

- *Information extraction.* Systems that extract factual knowledge from text must be trained with detailed annotations of documents. Users highlight entities or relations of interest, such as person and organization names, or whether a person works for a particular organization. Locating entities and relations can take a half-hour or more for even simple newswire stories (Settles et al., 2008a). Annotations for specific knowledge domains may require additional expertise, e.g., annotating gene and disease mentions in biomedical text usually requires PhD-level biologists.

- *Computational Biology.* Increasingly, machine learning is used to interpret and make predictions about data from the natural sciences, particularly biology. For example, biochemists can induce models that help explain and predict enzyme activity from hundreds of synthesized peptide chains (Smith et al., 2011). However, there are 20^n possible peptides of length n, which for

8-mers yields $20^8 \approx 2.6$ billion possibilities to synthesize and test. In practice, scientists might resort to random sequences, or cherry-picking subsequences of possibly interesting proteins, with no guarantee that either will provide much information about the chemical activity in question.

In all of these examples, data collection (for traditional supervised learning methods) comes with a hefty price tag, in terms of human effort and/or laboratory materials. If an active learning system is allowed to be part of the data collection process — to be "curious" about the task, if you will — the goal is to learn the task better with less training.

While the binary search strategy described in the previous section is a useful introduction to active learning, it is not directly applicable to most problems. For example, what if fruit safety is related not only to shape, but to size, color, and texture as well? Now we have four features to describe the input space instead of just one, and the simple binary search mechanism no longer works in these higher-dimensional spaces. Also consider that the bodies of different people might respond slightly differently to the same fruit, which introduces ambiguity or noise into the observations we use as labels. Most interesting real-world applications, like the ones in the list above, involve learning with hundreds or thousands of features (input dimensions), and the labels are often not 100% reliable.

The rest of this book is about the various ways we can apply the principles of active learning to machine learning problems in general. We focus primarily on classification, but touch on methods that apply to regression and structured prediction as well. Chapters 2–5 present, in detail, several *query selection* frameworks, or utility measures that can be used to decide which query the learner should ask next. Chapter 6 presents a unified view of these different query frameworks, and briefly touches on some theoretical guarantees for active learning. Chapter 7 summarizes the strengths and weaknesses of different active learning methods, as well as some practical considerations and a survey of more recent developments, with an eye toward the future of the field.

1.3   SCENARIOS FOR ACTIVE LEARNING

Before diving into query selection algorithms, it is worth discussing scenarios in which active learning may (or may not) be appropriate, and the different ways in which queries might be generated. In some applications, instance labels come at little or no cost, such as the "spam" flag you mark on unwanted email messages, or the five-star rating you might give to films on a social networking website. Learning systems use these flags and ratings to better filter your junk email and suggest new movies you might enjoy.

In cases like this, you probably have other incentives for providing these labels — like keeping your inbox or online movie library organized — so you provide many such labels "for free." Deploying an active learning system to carefully select queries in these cases may require significant engineering overhead, with little or no gains in predictive accuracy. Also, when only a relatively small number (e.g., tens or hundreds) of labeled instances are needed to train an accurate model, it may not be appropriate to use active learning. The expense of implementing the query framework might be greater than merely collecting a handful of labeled instances, which might be sufficient.

Active learning is most appropriate when the (unlabeled) data instances themselves are numerous, can be easily collected or synthesized, and you anticipate having to label many of them to train an accurate system. It is also generally assumed that the oracle answers queries about instance labels, and that the appropriate hypothesis class for the problem is more or less already decided upon (naive Bayes, decision trees, neural networks, etc.). These last two assumptions do not always hold, but for now let us assume that queries take the form of unlabeled instances, and that the hypothesis class is known and fixed[2]. Given that active learning is appropriate, there are several different specific ways in which the learner may be able to ask queries. The main scenarios that have been considered in the literature are (1) query synthesis, (2) stream-based selective sampling, and (3) pool-based sampling.

Query Synthesis. One of the first active learning scenarios to be investigated is learning with *membership queries* (Angluin, 1988). In this setting, the learner may request "label membership" for any unlabeled data instance in the input space, including queries that the learner synthesizes *de novo*. The only assumption is that the learner has a definition of the input space (i.e., the feature dimensions and ranges) available to it. Figure 1.4(a) illustrates the active learning cycle for the query synthesis scenario. Efficient query synthesis is often tractable and efficient for finite problem domains (Angluin, 2001). The idea of synthesizing queries has also been extended to regression learning tasks, such as learning to predict the absolute coordinates of a robot hand given the joint angles of its mechanical arm as inputs (Cohn et al., 1996). Here the robot decides which joint configuration to test next, and executes a sequence of movements to reach that configuration, obtaining resulting coordinates that can be used as a training signal.

Query synthesis is reasonable for some problems, but labeling such arbitrary instances can be awkward and sometimes problematic. For example, Lang and Baum (1992) employed membership query learning with human oracles to train a neural network to classify handwritten characters. They encountered an unexpected problem: many of the query images generated by the learner contained no recognizable symbols; these were merely artificial hybrid characters with little or no natural semantic meaning. See Figure 1.4(b) for a few examples: is the image in the upper-right hand corner a 5, an 8, or a 9? It stands to reason that this ambiguous image could help the learner discriminate among the different characters, if people were able to discriminate among them as well. Similarly, one could imagine query synthesis for natural language tasks creating streams of text or audio speech that amount to gibberish. The problem is that the data-generating distribution is not necessarily taken into account (and may not even be known), so an active learner runs the risk of querying arbitrary instances devoid of meaning. The stream-based and pool-based scenarios (described shortly) have been proposed to address these limitations.

---

[2]Relaxations of these and other assumptions are discussed in Chapter 7.

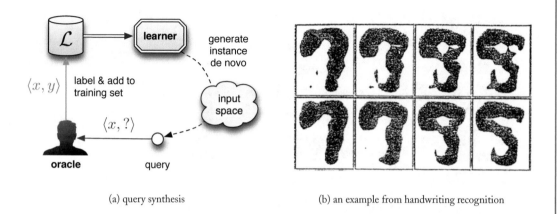

(a) query synthesis   (b) an example from handwriting recognition

Figure 1.4: (a) An active learner might synthesize query instances de novo. (b) Query synthesis can result is awkward and uninterpretable queries, such as these images generated by a neural network attempting to learn how to recognize handwritten digits. *Source*: Lang and Baum (1992), reprinted with kind permission of the authors.

Nevertheless, King et al. (2004, 2009) found a promising real-world application of query synthesis. They employ a "robot scientist" which executes autonomous biological experiments to discover metabolic pathways in the yeast *Saccharomyces cerevisiae*. Here, an instance corresponds to a mixture of chemical solutions that constitute a growth medium crossed with a particular yeast mutant. A label, then, is whether or not the mutant thrived in the growth medium. Experiments are chosen using an active learning approach based on inductive logic programming, and physically performed using a laboratory robot. The active approach results in a three-fold decrease in the cost of experimental materials compared to naively running the least expensive experiment, and a 100-fold decrease in cost compared to randomly chosen experiments. In this task, all query instances correspond to well-defined and meaningful experiments for the robot (or even a human) to perform. In other situations, arbitrary queries may be meaningless and nonsensical and thus difficult for an oracle to make judgements about.

Stream-Based Selective Sampling. An alternative to synthesizing queries is *selective sampling* (Atlas et al., 1990; Cohn et al., 1994). The key assumption is that obtaining an unlabeled instance is free (or inexpensive), so it can first be sampled from the actual distribution, and then the learner can decide whether or not to request its label. The approach is illustrated in Figure 1.5(a). This is sometimes called *stream-based* active learning, as each unlabeled instance is typically drawn one at a time from the input source, and the learner must decide whether to query or discard it. If the input distribution is uniform, selective sampling may not offer any particular advantages over

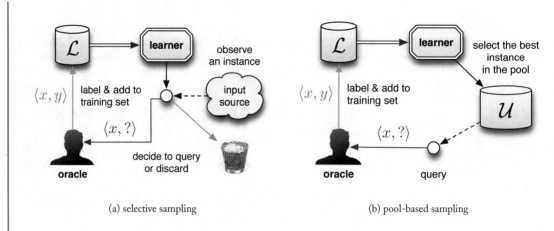

(a) selective sampling   (b) pool-based sampling

Figure 1.5: (a) In selective sampling, the learner decides whether to query or discard items from a stream of input instances. (b) Pool-based active learning queries the most informative instance from a large pool of available unlabeled data $\mathcal{U}$.

query synthesis. However, if the distribution is non-uniform and (more importantly) unknown, we are guaranteed that queries will still be sensible, since they come from a real underlying distribution.

The decision whether to query or discard an instance can be framed several ways. One approach is to define a measure of utility or information content (several such measures are discussed in Chapters 2–5) and make a biased random decision, such that instances with higher utility are more likely to be queried (c.f., Dagan and Engelson, 1995). Another approach is to compute an explicit *region of uncertainty* (Cohn et al., 1994), i.e., the part of the instance space that is still ambiguous to the learner, and only query instances that fall within it. A naive way of doing this is to set a minimum threshold on a utility measure which defines the region. Instances whose evaluation is above this threshold are then queried. Another somewhat more principled approach is to define the region that is still unknown to the overall model class, i.e., to the set of hypotheses consistent with the current labeled training set called the *version space* (Mitchell, 1982). In other words, if any two models of the same model class (but different parameter settings) agree on all the labeled data, but disagree on an unlabeled instance sampled from the data source, then that instance lies within the region of uncertainty. Calculating this region completely and explicitly is computationally expensive, however, and it must be maintained after each new query. As a result, approximations are used in practice (more details in Chapter 3).

The stream-based scenario has been studied in several real-world tasks, including part-of-speech tagging (Dagan and Engelson, 1995), sensor scheduling (Krishnamurthy, 2002), and learning ranking functions for information retrieval (Yu, 2005). Fujii et al. (1998) employed selective sampling for active learning in word sense disambiguation, e.g., determining if the word "bank" means land alongside a river or a financial institution in a given context (except they study Japanese words in their work). The approach not only reduces annotation effort, but also limits the size of the database used in nearest-neighbor learning, which in turn expedites the classification algorithm.

It is worth noting that some authors (e.g., Moskovitch et al., 2007; Thompson et al., 1999) use the term "selective sampling" to refer to the pool-based scenario described next. Under this interpretation, the term merely signifies that queries are made with a selected subset of instances sampled from a real data distribution. However, in most of the literature selective sampling refers to the stream-based scenario described here.

Pool-Based Sampling. For many real-world learning problems, large collections of unlabeled data can be gathered at once. This motivates *pool-based sampling* (Lewis and Gale, 1994), which assumes that there is a small set of labeled data $\mathcal{L}$ and a large pool of unlabeled data $\mathcal{U}$ available. The approach is illustrated in Figure 1.5(b). Queries are selected from the pool, which is usually assumed to be closed (i.e., static or non-changing), although this is not strictly necessary. Queries are typically chosen in a greedy fashion, according to a utility measure used to evaluate all instances in the pool (or, perhaps if $\mathcal{U}$ is very large, a subsample thereof). The binary search algorithm for the alien fruits example in Section 1.1 is a pool-based active learning algorithm.

The pool-based scenario has been studied for many real-world problem domains in machine learning, such as text classification (Hoi et al., 2006a; Lewis and Gale, 1994; McCallum and Nigam, 1998; Tong and Koller, 2000), information extraction (Settles and Craven, 2008; Thompson et al., 1999), image classification and retrieval (Tong and Chang, 2001; Zhang and Chen, 2002), video classification and retrieval (Hauptmann et al., 2006; Yan et al., 2003), speech recognition (Tür et al., 2005), and cancer diagnosis (Liu, 2004), to name only a few. In fact, pool-based sampling appears to be the most popular scenario for applied research in active learning, whereas query synthesis and stream-based selective sampling are more common in the theoretical literature.

The main difference between stream-based and pool-based active learning is that the former obtains one instance at a time, sequentially from some streaming data source (or by scanning through the data) and makes each query decision individually. Pool-based active learning, on the other hand, evaluates and ranks the entire collection of unlabeled data before selecting the best query. While the pool-based scenario appears to be much more common among application papers, one can imagine settings where the stream-based approach is more appropriate. For example, when memory or processing power is limited, as with mobile and embedded devices, or when the data set is too large to load into memory and must be scanned sequentially from disk. Unless otherwise noted, however, we will assume a pool-based scenario for our discussion of the algorithms discussed in the remainder of this book.

CHAPTER 2

Uncertainty Sampling

"Information is the resolution of uncertainty."

— *Claude Shannon*

2.1 PUSHING THE BOUNDARIES

Let us revisit the alien fruits problem from Section 1.1, and use it as a running example. Recall that we want to efficiently test fruits for ⊕ safe vs. ⊖ noxious to eat. One solution is to lay out all the fruits in a line from most round to most irregular, and use a binary search algorithm to actively select which fruits to test. This approach is fine for our simple thought experiment, but we want a more general search algorithm for arbitrary problems with many input dimensions, potentially many choices for output (e.g., multiple class labels or output structures), and probably even noisy training labels. As a starting point, recall that we can use supervised learning to select a threshold parameter θ that lies somewhere in the transition from one label to the other.

One reasonable way to choose this threshold value is to be as non-committal as possible: set θ to be halfway between the known ⊕ and ⊖ fruits which are closest together (this is what so-called *max-margin* learning algorithms do). Such a classifier should be fairly confident about its predictions for fruits which are far away from the thresholded classification boundary, but as x (the fruit's irregularity measure) approaches θ the model becomes much less certain. Intuitively, the instances that are least certain would offer the most information about the problem, since the more confident classifications are probably correct. What if we adopt a simple active learning strategy that queries the instance closest to the decision boundary? In fact, we would recover the binary search algorithm from Section 1.1, as illustrated by Figure 2.1.

This type of active learning strategy is commonly known as *uncertainty sampling* (Lewis and Catlett, 1994). The basic premise is that the learner can avoid querying the instances it is already confident about, and focus its attention instead on the unlabeled instances it finds confusing. The example in Figure 2.1 quantifies uncertainty by $|\theta - x|$, the distance of instance x from the boundary θ, which is a fine measure for hypothesis classes that provide such a distance measure. However, we are interested in generalizing active learning to more complex problems that go beyond binary classification in a relatively noise-free environment like this one. An elegant way to extend the approach is to use a probabilistic classifier which can output a posterior distribution $P_\theta(Y|x)$ over the label variable Y given the input and learned model parameters. Under this interpretation, we would want to query the instance for which $P_\theta(\hat{y}|x)$ — where $\hat{y}$ refers to the classifier's most likely

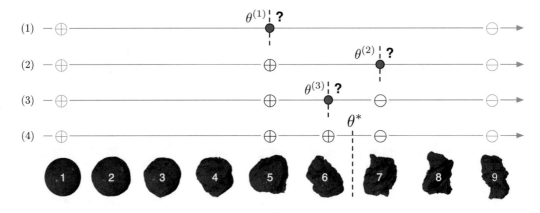

Figure 2.1: The binary search from Figure 1.3, re-interpreted as an uncertainty sampling approach. The best instance to query is deemed to be the one closest to the threshold θ.

prediction for x — is closest to a uniform distribution (0.5 in the case of binary classification). While a probabilistic interpretation is not strictly necessary, there has been significant work in machine learning on probabilistic classifiers, and graphical models in particular (for a thorough overview, see Koller and Friedman, 2009). By framing our discussion of uncertainty sampling in the language of probability, we can easily generalize the techniques in this chapter to a variety of interesting cases, including problems with many input features, multiple output labels, and even structured prediction tasks, which we will discuss later in this chapter.

2.2   AN EXAMPLE

To visualize the way in which uncertainty sampling generalizes to a noisy, two-dimensional classification problem, consider Figure 2.2. Figure 2.2(a) shows a toy data set constructed from two Gaussians centered at (-2,0) and (2,0) with standard deviation $\sigma = 1$. There are 400 instances total, 200 drawn from each class distribution. In a real-world setting, these instances may be available but their labels would not. Figure 2.2(b) illustrates the traditional supervised learning approach of randomly selecting instances for labeling. The line shows the linear decision boundary of a logistic regression model (i.e., where the posterior label probability equals 0.5) trained using 30 points. Notice that most of the labeled instances in this training set are far from zero on the horizontal axis, which is where the Bayes optimal decision boundary should be. As a result, this classifier only achieves 70% accuracy on the remaining unlabeled data. Figure 2.2(c) tells a different story, however: the active learner uses uncertainty sampling to focus on the instances closest to its decision boundary, assuming it can adequately explain the data in other parts of the input space. As a result, it avoids requesting labels for redundant or irrelevant instances, and achieves 90% accuracy using the same budget of 30 labeled instances.

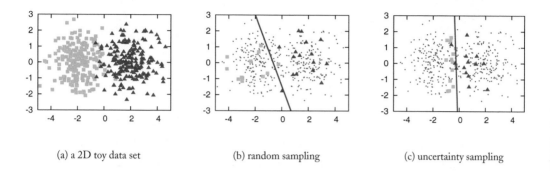

(a) a 2D toy data set   (b) random sampling   (c) uncertainty sampling

Figure 2.2: Uncertainty sampling with a toy data set. (a) 400 instances, evenly sampled from two class Gaussians. Instances are represented as points in a 2D input space. (b) A logistic regression model trained with 30 labeled instances randomly drawn from the problem domain. The line represents the decision boundary of the classifier. (c) A logistic regression model trained with 30 actively queried instances using uncertainty sampling.

1: $\mathcal{U} =$ a pool of unlabeled instances $\{x^{(u)}\}_{u=1}^{U}$
2: $\mathcal{L} =$ set of initial labeled instances $\{\langle x, y \rangle^{(l)}\}_{l=1}^{L}$
3: **for** $t = 1, 2, \ldots$ **do**
4:      $\theta = \textbf{train}(\mathcal{L})$
5:      select $x^* \in \mathcal{U}$, the most uncertain instance according to model θ
6:      query the oracle to obtain label y^*
7:      add $\langle x^*, y^* \rangle$ to $\mathcal{L}$
8:      remove x^* from $\mathcal{U}$
9: **end for**

Figure 2.3: Generic pool-based uncertainty sampling algorithm.

2.3   MEASURES OF UNCERTAINTY

A general active learning algorithm is presented in Figure 2.3. The key component of the algorithm with respect to designing an active learning system is line 5, and we need a way to measure the uncertainty of candidate queries in the pool. For binary classification, the "closest to the decision boundary (probability ≈ 0.5)" heuristic will suffice. But when we deal with problems and models that have posterior distributions over three or more labels — or even multiple output structures — we need a more general measure of uncertainty or information content. From this point on, let x_A^* denote the best instance that the utility measure A would select for querying.

Least Confident. A basic uncertainty sampling strategy is to query the instance whose predicted output is the least confident:

$$x^*_{LC} = \underset{x}{\text{argmin}}\ P_\theta(\hat{y}|x) \qquad (2.1)$$

$$= \underset{x}{\text{argmax}}\ 1 - P_\theta(\hat{y}|x),$$

where $\hat{y} = \text{argmax}_y\ P_\theta(y|x)$, the prediction with the highest posterior probability under the model θ. In other words, this strategy prefers the instance whose most likely labeling is actually the least likely among the unlabeled instances available for querying. One way to interpret this uncertainty measure is the expected 0/1-loss, i.e., the model's belief that it has mislabeled x. A drawback of this strategy is that it only considers information about the best prediction. Thus, it effectively throws away information about the rest of the posterior distribution.

Margin. A different active learning strategy is based on the output *margin*:

$$x^*_M = \underset{x}{\text{argmin}}\ \left[P_\theta(\hat{y}_1|x) - P_\theta(\hat{y}_2|x) \right] \qquad (2.2)$$

$$= \underset{x}{\text{argmax}}\ \left[P_\theta(\hat{y}_2|x) - P_\theta(\hat{y}_1|x) \right],$$

where $\hat{y}_1$ and $\hat{y}_2$ are the first and second most likely predictions under the model, respectively. Margin sampling addresses a shortcoming of the least confident strategy by incorporating the second best labeling in its assessment. Intuitively, instances with large margins are easy, since the learner has little doubt in how to differentiate between the two most probable alternatives. Instances with small margins are more ambiguous, though, and knowing the true labeling should help the model discriminate more effectively between them. However, for problems with a very large number of alternatives, the margin approach still ignores much of the output distribution.

Entropy. Perhaps the most general (and most common) uncertainty sampling strategy uses *entropy* (Shannon, 1948), usually denoted by H, as the utility measure:

$$x^*_H = \underset{x}{\text{argmax}}\ H_\theta(Y|x) \qquad (2.3)$$

$$= \underset{x}{\text{argmax}}\ -\sum_y P_\theta(y|x) \log P_\theta(y|x),$$

where y ranges over all possible labelings of x. Entropy is a measure of a variable's average information content. As such, it is often thought of as an uncertainty or impurity measure in machine learning. One interpretation of this strategy is the expected log-loss, i.e., the expected number of bits it will take to "encode" the model's posterior label distribution.

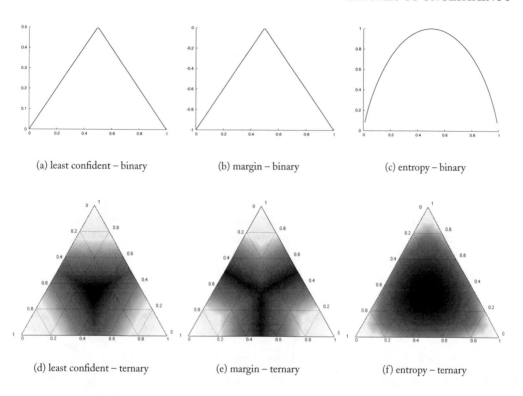

(a) least confident – binary

(b) margin – binary

(c) entropy – binary

(d) least confident – ternary

(e) margin – ternary

(f) entropy – ternary

Figure 2.4: Illustrations of various uncertainty measures. (a–c) Binary classification tasks. Each plot shows the utility score as a function of $P_\theta(\oplus|x)$, which is the posterior probability of the positive class. (d–f) Ternary classification tasks (three labels). Heatmap corners represent posterior distributions where one label is very likely, with the opposite edge plotting the probability range between the *other* two classes (when that label has low probability). The center of each heatmap is the uniform distribution.

Figure 2.4 visualizes the implicit relationship among these uncertainty measures. For binary classification (top row), all three strategies are monotonic functions of one another. They are symmetric with a peak about $P_\theta(\oplus|x) = 0.5$. In effect, they all reduce to querying the instance that is closest to the decision boundary. For a three-label classification task (bottom row), the relationship begins to change. For all three measures, the *most* informative instance lies at the center of the triangular simplex, because this represents where the posterior label distribution is most uniform (and therefore most uncertain under the model). Similarly, the *least* informative instances are at the three corners, where one of the classes has extremely high probability (and thus little model uncertainty). The main differences lie in the rest of the probability space. For example, the entropy measure does not particularly favor instances for which only one of the labels is highly *un*likely (i.e., along the outer side edges of the simplex), because the model is fairly certain that it is *not* the true label. The least confident and margin measures, on the other hand, consider such instances to be useful if the model has trouble distinguishing between the remaining two classes (i.e., at the midpoint of an outer edge). Intuitively, entropy seems appropriate if the objective function is to minimize log-loss, while the other two (particularly margin) are more appropriate if we aim to reduce classification error, since they prefer instances that would help the model better discriminate among specific classes.

2.4   BEYOND CLASSIFICATION

Not all machine learning is classification. In particular, we might want to use active learning to reduce the cost of training a model that predicts structured output, such as label sequences or trees. For example, Figure 2.5 illustrates how information extraction — the task of automatically extracting structured information like database entries from unstructured text — is framed as a sequence-labeling task. Let $\mathbf{x} = \langle x_1, \ldots, x_T \rangle$ be an observation sequence of length T with a corresponding label sequence $\mathbf{y} = \langle y_1, \ldots, y_T \rangle$. Words in a sentence correspond to *tokens* in $\mathbf{x}$, which are mapped to labels in $\mathbf{y}$.

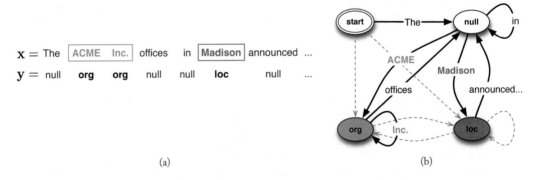

(a)   (b)

Figure 2.5: An information extraction example viewed as a sequence labeling task. (a) A sample input sequence $\mathbf{x}$ and corresponding label sequence $\mathbf{y}$. (b) A probabilistic sequence model represented as a finite state machine, illustrating the path of $\langle \mathbf{x}, \mathbf{y} \rangle$ through the model.

Figure 2.5(a) presents an example $\langle \mathbf{x}, \mathbf{y} \rangle$ pair. The labels indicate whether a given word belongs to an entity class of interest (org and loc in this case, for "organization" and "location," respectively) or null otherwise. Unlike simple classification, $\mathbf{x}$ is not represented by a single feature vector, but rather a sequence of feature vectors: one for each token (i.e., word). One approach is to treat each token as an instance, and train a classifier that scans through the input sequence, assigning output labels to tokens independently. However, the word "Madison," devoid of context, might refer to an location (city), organization (university), or even a person. For tasks such as this, *sequence models* based on probabilistic finite state machines, such as hidden Markov models or linear-chain conditional random fields, are considered the state of the art. An example sequence model is shown in Figure 2.5(b).

Such models can produce a probability distribution for every possible label sequence $\mathbf{y}$, the number of which can grow exponentially in the sequence length T.

Fortunately, uncertainty sampling generalizes fairly easily to probabilistic structured prediction models. For example, the least confident strategy is popular for information extraction using sequences (Culotta and McCallum, 2005; Settles and Craven, 2008), because the most likely output sequence $\hat{\mathbf{y}}$ and the associated $P_\theta(\hat{\mathbf{y}}|\mathbf{x})$ can be efficiently computed with dynamic programming. Selecting the best query is generally no more complicated or expensive than the standard inference procedure. The Viterbi algorithm (Corman et al., 1992) requires $O(TM)$ time, for example, where T is the sequence length and M is the number of label states. It is often possible to perform "N-best" inference using a beam search as well (Schwartz and Chow, 1990), which finds the N most likely output structures under the model. This makes it simple to compute the necessary probabilities for $\hat{\mathbf{y}}_1$ and $\hat{\mathbf{y}}_2$ in the margin strategy, and comes at little extra computational expense: the complexity is $O(TMN)$ for sequences, which for $N = 2$ merely doubles the runtime compared to the least confident strategy. Dynamic programs have also been developed to compute the entropy over all possible sequences (Mann and McCallum, 2007) or trees (Hwa, 2004), although this approach is significantly more expensive. The fastest entropy algorithm for sequence models requires $O(TM^2)$ time, which can be very slow when the number of label states is large. Furthermore, some structured models are so complex that they require approximate inference techniques, such as loopy belief propagation or Markov chain Monte Carlo (Koller and Friedman, 2009). In such cases, the least confident strategy is still straightforward since only the "best" prediction needs to be evaluated. However, the margin and entropy heuristics cease to be tractable and exact for these more complex models.

So far we have only discussed problems with discrete outputs — classification and structured prediction. Uncertainty sampling is also applicable to *regression*, i.e., learning tasks with continuous output variables. In this setting, the learner can simply query the unlabeled instance for which the learner has the highest output variance in its prediction. Under a Gaussian assumption, the entropy of a random variable is a monotonic function of its variance, so this approach is much in same the spirit as entropy-based uncertainty sampling. Another interpretation of variance is the expected squared-loss of the model's prediction. Closed-form estimates of variance can be computed for a variety of model classes, although they can require complex and expensive computations.

Figure 2.6 illustrates variance-based uncertainty sampling using an artificial neural network. The target function is a Gaussian in the range [-10,10], shown by the solid red line in the top row of plots. The network in this example has one hidden layer of three logistic units, and one linear output unit. The network is initially trained with two labeled instances drawn at random (upper left plot), and its variance estimate (lower left plot) is used to select the first query. This process repeats for a few iterations, and after three queries the network can approximate the target function fairly well. In general, though, estimating the output variance is nontrivial and will depend on the type of model being used. This is in contrast to the utility measures in Section 2.3 for discrete outputs, which only require that the learner produce probability estimates. In Section 3.4, we will discuss active learning approaches using ensembles as a simpler way to estimate output variance. Active learning

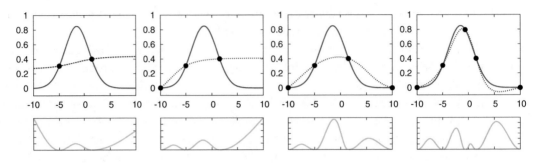

Figure 2.6: Variance-based uncertainty sampling for a toy 1D regression task. Each column represents an iteration of active learning. In the top row, solid lines show the target function to be learned, while dashed lines show a neural network approximation based on available training data (black dots). The bottom row plots the network's output variance across the input range, which is used to select the query for the next iteration.

for regression has a long history in the statistics literature, generally referred to as *optimal experimental design* (Federov, 1972). However, the statistics community generally eschews uncertainty sampling in lieu of more sophisticated strategies, which we will explore further in Chapter 4.

2.5   DISCUSSION

Uncertainty sampling is possibly the most popular active learning strategy in practice. Perhaps this is because of its intuitive appeal combined with the ease of implementation. Most of the common uncertainty-based utility measures do not require significant engineering overhead to use. In fact, as long as the learner can provide a confidence or probability score along with its predictions, any of the measures in Section 2.3 can be employed with the learner as a "black box." Standard classification or inference procedures can be used, leaving the choice of learning algorithm fairly modular. This is not necessarily the case for all active learning approaches. Furthermore, if inference is fast and tractable, then querying should also be fast and tractable.

Our discussion of uncertainty sampling has, thus far, been limited to the pool-based setting where the learner selects the "best" query from the set of unlabeled data $\mathcal{U}$. Uncertainty sampling can also be employed in the stream-based selective sampling setting, where unlabeled instances are drawn $x \backsim \mathcal{D}_X$ one at a time from an input distribution, and the learner decides on the spot whether to query or discard it. The simplest way to implement uncertainty-based selective sampling is to set a threshold on the uncertainty measure and use this to define a *region of uncertainty*. An instance is queried if it falls within the region of uncertainty, and discarded otherwise. The learner is re-trained after each new query instance (or batch of instances) is labeled and added to $\mathcal{L}$.

Figure 2.7 illustrates the idea of selective sampling using a neural network for a toy classification task. Positive instances lie inside the black box, and the network in this example has one

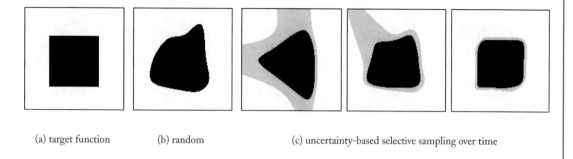

(a) target function   (b) random   (c) uncertainty-based selective sampling over time

Figure 2.7: Stream-based uncertainty sampling for a simple toy classification task. (a) Positive instances lie inside the black box in 2D. (b) After 100 random samples, the function learned by a neural network is still somewhat amorphous. (c) Uncertainty-based selective sampling at 20, 60, and 100 queries. The highlighted areas represent the region of uncertainty, which gradually shrinks and becomes more focused as the network grows more confident. The output of the resulting network after 100 queries is much more square-like than (b).

hidden layer of eight units. If we obtain labels for the first 100 instances drawn randomly from the 2D plane (i.e., passive learning), then the network tends to learn a nebulous, vaguely square-like shape like the one in Figure 2.7(b). Instead, we can define a region of uncertainty as any point on the plane where the network confidence is below some threshold $P_\theta(\hat{y}|x) < 0.9$. By limiting queries to the streaming instances within the region of uncertainty, the network learns the square shape much faster; Figure 2.7(c) shows a progression of the learned function at various intervals. The active network in this example discarded more than 1,000 "uninformative" instances before obtaining a training set of size 100 and learning the square function pictured above.

However, uncertainty sampling is not without problems. For one thing, the speed and simplicity of the approach comes at the expense of being shortsighted, as the utility scores are based on the output of a single hypothesis. To make matters worse, that hypothesis is often trained with very little data — given that a major motivating use case for active learning is when little to no labeled data exist — and that data is also inherently biased by the active sampling strategy. Thus, the learner's grasp of the problem is probably fairly myopic[1]. Figure 2.8 makes this point a bit more concrete. Here, the function to be learned is a pair of triangles in 2D, and the learner is again a two-layer neural network with eight hidden units. If we are unlucky, the small sample in Figure 2.8(b) used to train the initial network does not contain many instances from the white (i.e., negative) stripe in between the two triangles. As a result, the network over-generalizes and becomes quite confident that regions in the center of the plane are positive. As active learning progresses, the network continues to discard any instances that come from between the two triangles, and starts to learn a

---

[1]Sampling bias is an inherent problem in nearly every active learning strategy, not just uncertainty sampling. Section 7.7 discusses some of the practical dangers of sampling bias beyond the myopia discussed here.

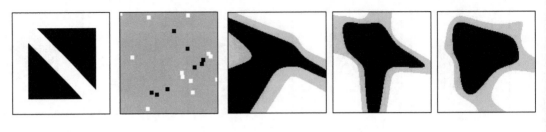

(a) target function   (b) initial sample   (c) uncertainty-based selective sampling over time

Figure 2.8: An example of uncertainty sampling failure. (a) Positive instances lie inside the two black triangles. (b) An initial random sample fails to draw many training instances from the negative space in between the triangles. (c) The trained network becomes overly confident that instances in the center are positive. As a result, it avoids that region and begins to learn a different, more square-like shape.

shape that is closer to a square-like "convex hull" of the two triangles instead. Even worse, one can imagine that the initial sample only contained positive instances from one of the two triangles, and the learner avoids querying in the region of the other triangle because it believes those instances are negative. While these may be pathological examples, the concerns are real, and uncertainty sampling has been observed to perform worse than random sampling in some applied settings (Schütze et al., 2006; Tomanek et al., 2009; Wallace et al., 2010a). The query strategy frameworks in the remainder of this book are aimed at alleviating or circumventing these issues.

CHAPTER 3

Searching Through the Hypothesis Space

"There are two possible outcomes: if the result confirms the hypothesis, then you've made a measurement. If the result is contrary to the hypothesis, then you've made a discovery."

— *Enrico Fermi*

3.1 THE VERSION SPACE

Recall that in machine learning parlance, a "hypothesis" is a particular computational model which attempts to generalize or explain the training data, and make predictions on new data instances. Let $\mathcal{H}$ denote the *hypothesis space*: the set of all hypotheses under consideration. For example, $\mathcal{H}$ could be comprised of all possible weights assigned to parameters in a perceptron or a given neural network structure, or all possible decision trees that use the input and output variables as decision nodes. A hypothesis, then, is a particular network configuration or decision tree in $\mathcal{H}$ that attempts to explain the training data or make accurate predictions on future instances as well as it can.

Now let $\mathcal{V} \subseteq \mathcal{H}$ denote the *version space* (Mitchell, 1982): the subset of hypotheses which are *consistent* with the training data. That is, they make correct predictions for all labeled training instances in $\mathcal{L}$. In some sense, the version space represents the candidate hypotheses that can explain the observed training data equally well. It is the task of a learning algorithm to choose which one of these hypotheses to use in characterizing the data or making future predictions. If we suppose that the underlying function to be learned is "separable" — that is, it can be perfectly expressed by one of these hypotheses — then as we obtain more labeled data, the area of the version space $|\mathcal{V}|$ becomes smaller and smaller. As a result, the set of hypotheses therein will approximate the underlying target function more and more accurately. This suggests that an active learning algorithm should try to obtain new training instances that will quickly minimize $|\mathcal{V}|$.

Consider again the alien fruits example from Section 1.1. The hypothesis space $\mathcal{H}$ consists of all thresholds we might draw to distinguish between safe $\oplus$ and noxious $\ominus$ fruits. The version space $\mathcal{V}$ is therefore the subset of these thresholds that agree with all the fruits that have been tested so far, i.e., the thresholds bounded by the two nearest fruits labeled $\oplus$ and $\ominus$. At each step in the active learning process, we want to pose a query that will reduce $|\mathcal{V}|$ as much as possible. For a binary

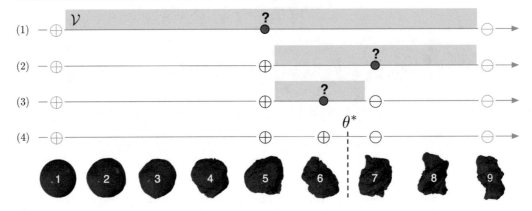

Figure 3.1: The binary search from Figure 1.3, re-interpreted as a search through the version space. The shaded box labeled $\mathcal{V}$ at each step represents the version space (the set of possible hypotheses which are consistent with the labeled instances). By posing queries that bisect $\mathcal{V}$, we efficiently search for good hypotheses by quickly eliminating the inconsistent ones.

classification problem, we can (ideally) cut $\mathcal{V}$ in half: if the oracle's answer to the query is $\oplus$, then we can discard all hypotheses that would label it $\ominus$ (and vice versa). To echo the sentiments of Fermi's epigram at the beginning of this chapter, we want the learner to pose queries that lead to as many "discoveries" as possible. As Figure 3.1 shows, this is precisely what the binary search algorithm does. Now the question becomes, "How can we generalize this notion to more interesting learning problems than the 1D threshold?"

3.2   UNCERTAINTY SAMPLING AS VERSION SPACE SEARCH

We want to conduct a directed search through the hypothesis space by testing unlabeled instances, in order to minimize the number of "legal" hypotheses under consideration ($\mathcal{V}$). Conceptually, this approach is in contrast to uncertainty sampling (Chapter 2), which makes query decisions based on the confidence of a *single* hypothesis. In special cases, however, it turns out that uncertainty sampling represents an attempt to incrementally reduce the size of the version space. After all, a binary search through a 1D input space is justified under both interpretations (Figures 2.1 and 3.1). This is because we used a so-called *max-margin* classifier, which positions itself in the center, halfway between positive and negative instances. Now let us consider *support vector machines* (SVMs) for binary classification in more complex problem representations, which have been the subject of much theoretical and empirical research in recent years. In its simplest form, an SVM is a $(K-1)$-dimensional hyperplane in a K-dimensional feature space $\mathcal{F}$. The hypothesis space $\mathcal{H}$ includes all such hyperplanes, and the version space $\mathcal{V}$ includes those which can correctly separate the training data in feature space. Interestingly, there exists a duality between the feature space $\mathcal{F}$ and the

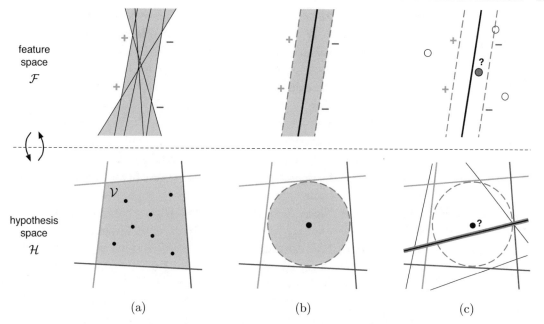

Figure 3.2: The version space duality. (a) Instance points in feature space $\mathcal{F}$ correspond to hyperplanes in hypothesis space $\mathcal{H}$, and vice versa. (b) SVMs attempt to maximize the margin (distance) between the decision boundary and the closest support vectors in $\mathcal{F}$, corresponding to a hypothesis at the center of the maximal bounded hypersphere in $\mathcal{H}$. (c) Querying the instance closest to the decision boundary in $\mathcal{F}$ approximately bisects the version space in $\mathcal{H}$.

hypothesis space $\mathcal{H}$: points in $\mathcal{F}$ correspond to hyperplanes (or boundaries) in $\mathcal{H}$ and vice versa[1]. Therefore, labeled data instances impose geometric constraints on $\mathcal{V}$ in the hypothesis space. This is easy to see in 1D (e.g., Figure 3.1): consistent hypotheses must lie between the two nearest labeled instances.

Figure 3.2 illustrates the so-called *version space duality* in 2D, and how it can be useful for active learning with SVMs. The top row depicts $\mathcal{F}$: data instances are points while hypotheses are lines in feature space. The bottom row is meant to depict the same information in $\mathcal{H}$: data instances are lines while hypotheses are points in the hypothesis space. The highlighted regions in Figure 3.2(a) show the *region of disagreement* among hypotheses in the version space $\mathcal{V} \subseteq \mathcal{H}$. SVMs are trained under the notion that the best separating hyperplane is the one that represents the largest separation, or *margin*, between the two classes, since larger margins commonly result in lower generalization error. Data instances on one side of the plane are classified as positive, while points on the other side are negative, and the training instances closest to the hyperplane (i.e., that lie right on the margin) are called *support vectors*. Figure 3.2(b) represents the max-margin hyperplane as a solid black line in

[1]See see Vapnik (1998) or Tong and Koller (2000) for a more technical exposition.

$\mathcal{F}$, and the margin boundaries as dashed grey lines. In $\mathcal{H}$, this hypothesis corresponds to the center of the largest hypersphere bounded by the labeled instance vectors. In other words, the best SVM hypothesis is the one that lies approximately in the center of the version space.

Uncertainty sampling involves querying the unlabeled instance that is closest to the decision boundary in $\mathcal{F}$ (in the case of binary classification, anyway). Since the SVM decision boundary should lie near the center of the version space $\mathcal{V}$, this heuristic corresponds to querying the instance vector in $\mathcal{H}$ that runs nearest to the center of $\mathcal{V}$, roughly bisecting it. The analogy is shown in Figure 3.2(c). By obtaining a label for this least confident instance vector, $|\mathcal{V}|$ should be cut approximately in half and the bounded hypersphere governing the decision boundary will be further constrained (regardless of its label), which is the desired outcome. There are several machine learning methods besides SVMs that attempt to maximize the margin, such as variants of boosting (Warmuth et al., 2008) and the voted perceptron (Freund and Schapire, 1999). Conceptually, uncertainty sampling is equally principled as a version-space active learning approach for these algorithms. Furthermore, margin-based active learning has been demonstrated to work well for several practical applications of large-margin classifiers (Jain et al., 2010; Schohn and Cohn, 2000; Tong and Koller, 2000).

However, there is still reason to be cautious about uncertainty sampling. For one thing, margin-based uncertainty sampling will only bisect the version space if it is symmetrical. In Figure 3.2(c), $\mathcal{V}$ is square-like and therefore approximately symmetrical. After labeling the query instance, however, it is no longer a nearly regular-shaped polygon. As a result, the largest bounded hypersphere in the revised version space will be far away from the center of $\mathcal{V}$, and the active learner might waste time winnowing away only a few fringe hypotheses at a time. To address these limitations, Tong and Koller (2000) proposed alternative algorithms for SVMs that try to anticipate the expected reduction in $|\mathcal{V}|$ in a decision-theoretic style analysis[2]. Finally, uncertainty sampling is commonly used with classifiers that do not attempt to maximize the margin. The interpretation of uncertainty sampling as a directed search through the version space does not necessarily carry over to these other methods, and there is no theoretical argument that it will work well.

3.3   QUERY BY DISAGREEMENT

One of the earliest active learning algorithms explicitly motivated by reducing the version space was proposed by Cohn et al. (1994), based on a simple but powerful idea. The algorithm, which we will call *query by disagreement* (QBD), is shown in Figure 3.3. QBD assumes the stream-based selective sampling scenario (see Section 1.3), where data come in a stream and the learner decides whether to query or discard them in real time. The approach essentially maintains the working version space $\mathcal{V}$, and if a new data instance x comes along for which any two legal hypotheses disagree, then x's labeling cannot be inferred and its true label should be queried. However, if all the legal hypotheses do agree, then the label can be inferred and x can be safely ignored. Geometrically, one can think about defining a *region of disagreement* $\mathrm{DIS}(\mathcal{V})$, and only querying instances from the stream that

---

[2]We will not go into detail on these alternative SVM algorithms in this book, but they are similar in spirit to the expected error reduction framework (Chapter 4), and can suffer from some of the same issues of computational complexity.

1: $\mathcal{V} \subseteq \mathcal{H}$ is the set of all "legal" hypotheses
2: **for** $t = 1, 2, \ldots$ **do**
3:      receive instance $x \backsim \mathcal{D}_X$
4:      **if** $h_1(x) \neq h_2(x)$ for any $h_1, h_2 \in \mathcal{V}$ **then**
5:          query label y for instance x
6:          $\mathcal{L} = \mathcal{L} \cup \langle x, y \rangle$
7:          $\mathcal{V} = \{h : h(x') = y' \text{ for all } \langle x', y' \rangle \in \mathcal{L}\}$
8:      **else**
9:          do nothing; discard x
10:     **end if**
11: **end for**
12: return the labeled set $\mathcal{L}$ for training

Figure 3.3: The query by disagreement algorithm (QBD).

fall within this region. For example, the highlighted region in the top row of Figure 3.2 is the region of disagreement for linear hyperplanes, given four labeled instances. This algorithm, based on a simple intuition of disagreement, can exponentially reduce the number of labeled instances needed to train an accurate classifier compared to traditional random i.i.d. sampling (the proof is presented in Section 6.2).

There are a few practical shortcomings of QBD as it is presented here. For one thing, the size of the version space may be infinite. This means that, in practice, the version space of an arbitrary hypothesis class cannot be explicitly represented in memory. In some special cases, it may be possible to bound $\mathcal{V}$ analytically, but not always. One solution to this problem is to store the version space *implicitly* instead. For example, instead of exhaustively comparing all pairs of hypotheses in line 4 of Figure 3.3, we could consider two speculative hypotheses:

$$
\begin{aligned}
h_1 &= \textbf{train}(\mathcal{L} \cup \langle x, \oplus \rangle), \text{ and} \\
h_2 &= \textbf{train}(\mathcal{L} \cup \langle x, \ominus \rangle),
\end{aligned}
$$

where $\oplus$ and $\ominus$ denote positive and negative labels, and $\textbf{train}(\cdot)$ is the learning algorithm we intend to use in active learning: it takes a labeled data set and returns a classifier consistent with the data (if one exists). The rest of the algorithm remains unchanged. If these two hypotheses disagree on how to label x, then it is queried, otherwise it is ignored. Note that for more than two classes, we would need to build a speculative classifier for each label, and then compare all of their predictions. However, if the training procedure is computationally expensive, this can lead to a drastic slowing down of the algorithm.

A different approach might be to exploit a partial ordering on the "generality" of hypotheses in $\mathcal{V}$. Consider this example of axis-parallel box classifiers:

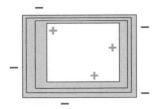

The smallest box, tightly bound around positive instances, is a very conservative hypothesis, labeling very few instances as positive. The other boxes are more liberal with the input regions they will classify as positive. Imagine that we have two subsets $\mathcal{S}, \mathcal{G} \subseteq \mathcal{V}$, where $\mathcal{S}$ contains the set of all "most specific" hypotheses, and $\mathcal{G}$ contains all the "most general" hypotheses in the version space (there may be multiple general hypotheses which are not necessarily comparable). Now we can use the QBD algorithm from Figure 3.3, where $h_1 \in \mathcal{S}$ and $h_2 \in \mathcal{G}$ in line 4, leaving the rest unchanged. This alleviates the problem of maintaining the entire version space since $\mathrm{DIS}(\mathcal{V})$ can be accurately summarized with these two sets. If a query falls within $\mathrm{DIS}(\mathcal{V})$ and is positive, then hypotheses in $\mathcal{S}$ must be made more general; likewise if it proves negative, those in $\mathcal{G}$ are made more specific. Unfortunately, maintenance of the $\mathcal{S}$ and $\mathcal{G}$ sets is still prohibitively expensive in general, as Haussler (1994) pointed out that they can grow exponentially in the size of $\mathcal{L}$. Another problem is that the data may not be perfectly separable after all (i.e., there is label noise), in which case the version space technically does not even exist.

However, this does bring to mind an approach that maintains only two hypotheses, h_S and h_G, which are *roughly* the most specific and most general hypotheses in $\mathcal{V}$, respectively. That is, they are very conservative and liberal hypotheses that aim to be consistent with $\mathcal{L}$. The region of disagreement can be approximated by the instances for which these two extremes disagree, $\mathrm{DIS}(\mathcal{V}) \approx \{x \in \mathcal{D}_X : h_S(x) \neq h_G(x)\}$. An simple way to encourage this specific/general behavior is to sample a set of "background" data points $\mathcal{B} \curvearrowleft \mathcal{D}_X$, and add them to $\mathcal{L}$ with artificial labels. To train the h_S model, these artificial data are given the $\ominus$ label, thus making it conservative in its predictions of the unlabeled data. Similarly, they are given the $\oplus$ label when training the h_G model, making it much more willing to label the unknown regions positive.

Figure 3.4 compares the $\mathcal{SG}$-based QBD approach to random sampling and uncertainty sampling for the simple two triangles learning task from Section 2.5. In addition to the initial random sample of 20 instances shown in Figure 3.4(b), 200 background instances are added when training the h_S and h_G networks. These background labels are also weighted by 1/200 so as not to overwhelm the actual labeled data in $\mathcal{L}$. Figure 3.4(f) shows how these two networks interact with one another as active learning progresses up to 100 labeled instances. The highlighted areas represent $\mathrm{DIS}(\mathcal{V})$, while black and white regions are where the networks agree on positive and negative labels, respectively. Figure 3.4(g) shows the analogous sequence using uncertainty sampling for a single best MAP hypothesis (using only $\mathcal{L}$ as training data). Notice that, since the initial sample does not

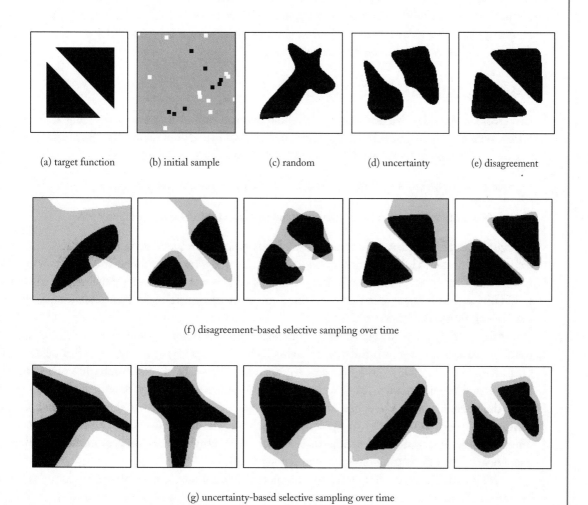

(a) target function   (b) initial sample   (c) random   (d) uncertainty   (e) disagreement

(f) disagreement-based selective sampling over time

(g) uncertainty-based selective sampling over time

Figure 3.4: Selective sampling for learning the toy two triangles task. (a) The target concept. (b) An initial random sample of 20 labeled instances. (c) The output of a neural network after 100 random samples. (d) A network after 100 instances queried using uncertainty sampling (see Section 2.5). (e) A network after 100 instances using the $\mathcal{SG}$-based QBD approach. (f) The regions of disagreement at 20, 40, 60, 80, and 100 instances queries using QBD. (g) The analogous regions of uncertainty using uncertainty sampling.

contain many instances in the center of the image, uncertainty sampling avoids that area until very late. In contrast, $\mathcal{SG}$-based QBD discovers that there are two triangle-like regions very early on. Figures 3.4(c–e) show the final networks trained on labeled data sets of size 100. While uncertainty sampling does a better job than random sampling at discovering the two different objects, QBD does a much better job at modeling them as triangles.

3.4   QUERY BY COMMITTEE

Disagreement-based active learning is compelling because, unlike uncertainty sampling, it attempts to pose queries that distinguish among different choices of hypotheses themselves. So far, the heuristics we have discussed have been based on two assumptions about the notion of disagreement:

- it is measured among all hypotheses $h \in \mathcal{V}$ (or two approximate extremes h_S and h_G);

- it is a binary measure, i.e., no controversial instance matters more than any other.

It might be helpful to relax these assumptions. First, it can be problematic to measure disagreement among all hypotheses in the version space, even if we imperfectly approximate it with two extremes h_S and h_G. Even worse, perhaps the data are noisy and $\mathcal{V}$ is not even properly defined. Second, we would like to use disagreement-based heuristics in a pool-based setting, i.e., when we want to query the "most informative" instance $x \in \mathcal{U}$.

The *query by committee* (QBC) algorithm relaxes these two assumptions to make disagreement-based active learning more broadly applicable. Technically speaking, the original formulation of QBC (Seung et al., 1992) and its subsequent analysis (Freund et al., 1997) sampled two random hypotheses from the version space at each iteration and used a binary notion of disagreement. Over the years, however, QBC has come to refer to any disagreement-based approach that uses a "committee" — or ensemble — of hypotheses which we will denote $\mathcal{C}$. These can take a variety of different forms in practice. All one needs is a method for obtaining hypotheses in the committee, and a heuristic for measuring disagreement among them.

If the data are noise-free and the version space is well defined, it is possible to randomly sample hypotheses from $\mathcal{V}$ with certain learning algorithms, such as the perceptron (Freund et al., 1997) and winnow (Liere and Tadepalli, 1997). Although it may be the case that this sampling strategy is computationally inefficient, and it does not work for noisy data anyway. There are several alternatives that get around these issues. One could take a Bayesian approach by sampling hypotheses from a posterior distribution $P(\theta|\mathcal{L})$. For example, McCallum and Nigam (1998) did this for naive Bayes by using the Dirichlet distribution over model parameters, whereas Dagan and Engelson (1995) sampled hidden Markov models by using the Normal distribution. It is also quite natural to think about using generic ensemble learning algorithms to construct a committee. For example, *query by boosting* uses a boosting algorithm (Freund and Schapire, 1997) to construct a sequence of hypotheses that gradually become more an more specialized, by increasingly focusing them on the erroneous instances in $\mathcal{L}$. Another approach, *query by bagging*, employs the bagging algorithm (Breiman, 1996) to train a committee of ensembles based on resamples of $\mathcal{L}$ with replacement, which is meant to smooth out high-variance predictions. These ensemble-based committee methods were proposed by Abe and Mamitsuka (1998) and have been used quite extensively since. Melville and Mooney (2004) proposed a different ensemble-based method that explicitly encourages diversity among the committee members. Muslea et al. (2000) constructed a committee of hypotheses by partitioning the feature space into conditionally independent committee-specific subsets, to be used in conjunction with semi-supervised learning (see Section 5.3). Regarding the number of committee members to use, there is no general rule of thumb. It appears from the literature that committees of five to fifteen hypotheses are quite common. However, even small committee sizes (e.g., two or three hypotheses) have been shown to work well.

There are also a variety of heuristics for measuring disagreement in classification tasks, but we will focus on two dominant trends. The first is essentially a committee-based generalization of uncertainty measures. For example, *vote entropy* is defined as:

$$x_{VE}^* = \operatorname*{argmax}_{x} - \sum_{y} \frac{\text{vote}_{\mathcal{C}}(y, x)}{|\mathcal{C}|} \log \frac{\text{vote}_{\mathcal{C}}(y, x)}{|\mathcal{C}|}, \tag{3.1}$$

where y ranges over all possible labelings, $\text{vote}_{\mathcal{C}}(y, x) = \sum_{\theta \in \mathcal{C}} \mathbf{1}_{\{h_\theta(x)=y\}}$ is the number of "votes" that the label y receives for x among the hypotheses in committee $\mathcal{C}$, and $|\mathcal{C}|$ is the committee size. This formulation is a "hard" vote entropy measure; we can also define a "soft" vote entropy which accounts for each committee member's confidence:

$$x_{SVE}^* = \operatorname*{argmax}_{x} - \sum_{y} P_{\mathcal{C}}(y|x) \log P_{\mathcal{C}}(y|x), \tag{3.2}$$

where $P_{\mathcal{C}}(y|x) = \frac{1}{|\mathcal{C}|} \sum_{\theta \in \mathcal{C}} P_\theta(y|x)$ is the average, or "consensus" probability that y is the correct label according to the committee. These disagreement measures are essentially Bayesian versions of entropy-based uncertainty sampling from Equation 2.3, using $\text{vote}(y, x)/|\mathcal{C}|$ or $P_{\mathcal{C}}(y|x)$ as the ensemble's posterior label estimate as opposed to that of a single hypothesis (or "point estimate"). Intuitively, this should smooth out any hard over-generalizations made by a single hypothesis. Analogous ensemble-based generalizations can be made for the least-confident and margin heuristics (Equations 2.1 and 2.2) as well.

A disagreement measure of a different flavor is based on *Kullback-Leibler (KL) divergence* (Kullback and Leibler, 1951), an information-theoretic measure of the difference between two probability distributions. In this case, we want to quantify disagreement as the average divergence of each committee member θ's prediction from that of the consensus $\mathcal{C}$:

$$x_{KL}^* = \operatorname*{argmax}_{x} \frac{1}{|\mathcal{C}|} \sum_{\theta \in \mathcal{C}} KL\big(P_\theta(Y|x) \,\|\, P_{\mathcal{C}}(Y|x) \big), \tag{3.3}$$

where KL divergence is defined to be:

$$KL\big(P_\theta(Y|x) \,\|\, P_{\mathcal{C}}(Y|x) \big) = \sum_{y} P_\theta(y|x) \log \frac{P_\theta(y|x)}{P_{\mathcal{C}}(y|x)}.$$

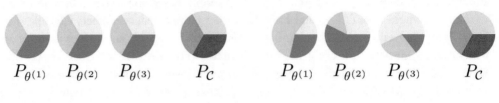

(a) uncertain but in agreement   (b) uncertain and in disagreement

Figure 3.5: Examples of committee and consensus distributions. $P_{\theta^{(i)}}$ refers the output distribution of the ith hypothesis, and P_C represents the consensus across all committee members.

There is a key conceptual difference in how vote entropy and KL divergence quantify disagreement. Consider the probability distributions in Figure 3.5(a). The output distributions for individual hypotheses in this example are relatively uniform, so the consensus output distribution is also uniform. Figure 3.5(b) paints a slightly different picture: the individual hypothesis distributions are non-uniform and each prefers a different label. But the consensus ends up being fairly uniform in this case as well. Vote entropy (3.2) only considers P_C and thus cannot distinguish between the two, since they both have high entropy in the consensus. In a way, then, vote entropy is like a Bayesian generalization of entropy-based uncertainty sampling, relying not on a single point estimate of model parameters, but aggregating uncertainties across a set of hypotheses. However, querying an instance with predictions like Figure 3.5(a) is not really in the disagreement-based spirit of QBC. The consensus label is uncertain, but this is because all committee members *agree* that it is uncertain. KL divergence (3.3), on the other hand, would favor an instance with predictions like Figure 3.5(b): the consensus here is uncertain, but the individual committee members vary wildly in their predictions, which is more in line with the notion of disagreement.

To make these intuitions a bit more concrete, consider active learning starting from the labeled data shown in Figure 3.6(a). Uncertainty sampling would train a single classifier on these data instances, and favor unlabeled instances that fall in less confident regions of the input space according to that single hypothesis. Figure 3.6(b) shows a heatmap of an entropy-based uncertainty sampling approach (2.3) using a single multinomial logistic regression classifier (also known as a maximum entropy classifier in the literature). Darker regions indicate higher utility according to the heuristic. Here we see that this single model is very confident in most of the input space, and as a result will prefer to only query instances that fall in a narrow region right along the decision boundary. Figure 3.6(c) shows the analogous heatmap for QBC using soft vote entropy (3.2), with a committee of ten classifiers obtained by the bagging algorithm (Breiman, 1996). The ensemble's preferences still take on a sort of 'Y' shape, but it is clearly smoother and less overly-confident about most of the input space. Figure 3.6(d) shows a heatmap for the KL divergence heuristic (3.3) using the exact same committee of ten classifiers. Even though the task, training data, and classifiers are all the same,

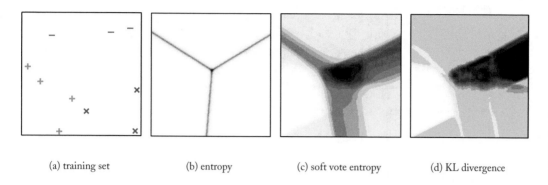

(a) training set   (b) entropy   (c) soft vote entropy   (d) KL divergence

Figure 3.6: Visualization of different active learning heuristics. (a) A small training set with three classes in 2D. (b) A heatmap of the input space according to entropy-based uncertainty sampling. Darker regions are deemed more informative by the MAP-estimate logistic regression classifier. (c) Soft vote entropy using a committee of ten logistic regression classifiers obtained by bagging. (d) KL divergence using the exact same committee of ten classifiers.

the two QBC heatmaps look very different: the KL divergence measure determines that the two leftmost decision boundaries, while uncertain, are not as contentious as the completely unknown region spanning from the center to the upper-right. In other words, the leftmost decision boundaries resemble posterior distributions like those Figure 3.5(a) — uncertain but in agreement — whereas the upper-right boundary is more similar to Figure 3.5(b) — uncertain and in disagreement.

Aside from these two dominant themes, a few other disagreement measures for QBC have been described in the literature for classification. One example is Jensen-Shannon divergence (Melville et al., 2005), which is essentially a symmetric and smoothed version of KL divergence. Ngai and Yarowsky (2000) also proposed an interesting measure called the F-compliment, which uses the well-known F_1-measure from the information retrieval literature (Manning and Schütze, 1999) to quantify (dis)agreement among committee members.

QBC can also be employed in regression settings, i.e., by measuring disagreement as the variance among committee members' predictions. The idea is illustrated in Figure 3.7 for learning a Gaussian target function (solid red line) in the range [-10,10] using bagged regression trees (dotted blue line). The ensemble of trees is initially trained with two labeled instances drawn at random (upper left plot), and the variance estimate (lower left plot) is used to select the first query. This process repeats and as more variance-based instances are queried, the function approximation improves. Burbidge et al. (2007) analyzed this rarely-used approach to active data selection for real-valued target functions, and found that it works well if the inductive bias of the learner and the label noise are both small. However, if the model class is misspecified or improperly biased (i.e., a linear regression being used to actively learn a nonlinear function) a QBC approach based on output variance can produce lower-quality models than randomly sampled data.

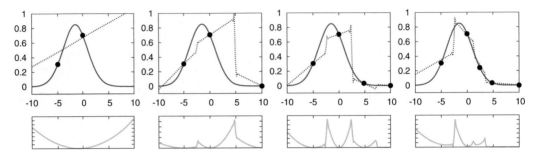

Figure 3.7: Variance-based QBC for a toy 1D task using bagged regression trees. Each column represents an iteration of active learning. In the top row, solid lines show the target function to be learned, while dashed lines show the mean output of an ensemble of ten regression trees induced from the training data (black dots). The bottom row plots the ensemble's output variance, which is used to select the query for the next iteration.

3.5   DISCUSSION

In this chapter, we have looked at a variety of active learning approaches that operate on a simple premise: select queries that eliminate as many "bad" hypotheses as possible. This is in contrast to the vanilla uncertainty sampling strategies from Chapter 2, which use a single hypothesis to select queries that simply appear to be confusing.

There is an information-theoretic interpretation of the subtleties at play here. Recall that we want to select a query x whose labeling would provide the most information about the hypotheses space (i.e., most reduces uncertainty about the choice of hypothesis itself). Let $I(Y; \mathcal{V})$ denote *mutual information* between the label variable Y and the (possibly noisy) version space $\mathcal{V}$, with some abuse of notation (namely, dropping dependence on x and treating $\mathcal{V}$ as a random variable). What we want, then, is to query the instance that maximizes $I(Y; \mathcal{V})$, providing the most mutual information between a query label and the version space. Consider the well-known information-theoretic identity:

$$I(Y; \mathcal{V}) = H(\mathcal{V}) - H(\mathcal{V}|Y)$$

$$= H(\mathcal{V}) - \mathbb{E}_Y[H(\mathcal{V}|y)], \tag{3.4}$$

where H denotes entropy. Assuming that $H(\mathcal{V}) \propto |\mathcal{V}|$ — both are measures of uncertainty about or complexity of the space of "good" hypotheses — then this justifies an approach that searches for queries that reduce $|\mathcal{V}|$ in expectation over all possible labelings. Queries that approximately bisect the version space can accomplish this, such as the examples early in this chapter: binary search in 1D, uncertainty sampling with max-margin classifiers like SVMs, or the QBD algorithm in its purest form. Certain query synthesis algorithms, which create new data instances de novo from the instances space, can be interpreted as generating queries to most constrain the version space in this way (Angluin, 1988; King et al., 2004).

However, an analysis that tries to maximize Equation 3.4 directly can be very complicated. Fortunately, mutual information is symmetric and there is an equivalent identity:

$$I(Y; \mathcal{V}) = H(Y) - H(Y|\mathcal{V})$$

$$= H(Y) - \mathbb{E}_{\theta \in \mathcal{V}}\big[H_\theta(Y)\big]. \qquad (3.5)$$

This states that we want to query instances which are uncertain — $H(Y)$ is high — but in such a way that there is disagreement among the competing hypotheses. If the choices of θ largely agree on y, then $\mathbb{E}_{\mathcal{V}}[H_\theta(Y)]$ will be very close to $H(Y)$, and the information content is deemed to be low; the more they disagree, the lower the second term is and the higher the overall information content. However, since the size of $\mathcal{V}$ can be intractably large in practice, the QBC algorithm uses a committee $\mathcal{C} \approx \mathcal{V}$ as a approximation to the full set of candidate hypotheses.

Mutual information, it turns out, is also equivalent to the KL divergence between the joint distribution $P(Y, \mathcal{V})$ and the product of the product of their marginal distributions:

$$I(Y; \mathcal{V}) = KL\big(P(Y, \mathcal{V}) \parallel P(Y)P(\mathcal{V})\big)$$

$$= \mathbb{E}_{\theta \in \mathcal{V}}\Big[KL\big(P_\theta(Y) \parallel P(Y)\big)\Big].$$

The second line follows from the fact that $P(y|\theta) = P(y, \theta)/P(\theta)$, so we can marginalize out the specific hypothesis θ. If we use a QBC approach, assume that each committee member is equally probable (i.e., $P(\theta) = 1/|\mathcal{C}|$), and approximate the marginal $P(Y) \approx P_\mathcal{C}(Y)$, then this is equivalent to the KL divergence utility measure defined in Equation 3.3. In other words, disagreement-based approaches are an attempt to maximize the mutual information between an instance's unknown label and the learning algorithm's unknown hypothesis. In contrast, entropy-based uncertainty sampling only attempts to maximize the entropy $H(Y) = I(Y; Y)$, or "self information" of an instance's unknown label. By doing this, uncertainty sampling does not try to obtain information about the hypothesis space itself, in general, whereas QBC and other disagreement-based active learning methods do.

However, methods based on uncertainty sampling and hypothesis search alike can suffer if our primary interest reducing generalization error on future unseen instances. To see why, consider a variant of our running alien fruits example. Imagine that there are two types of fruits that come in different colors: red and grey. They both have the property that smooth fruits are safe and irregular fruits are noxious, but they have different thresholds. Thus, our hypotheses now consist of two parameters, θ_1 and θ_2, one for each color fruit. Figure 3.8 shows how this might be problematic for both uncertainty sampling and methods that try to reduce the version space. There are two candidate queries that will cut the size of the version space in half: A, which is a grey fruit, and B, which is red. As a result, they both look equally informative. But the impact that each query will have on future classification error depends on the distribution of colored fruits. If that distribution is fairly balanced and uniform across both red and grey dimensions, then B is probably a more informative

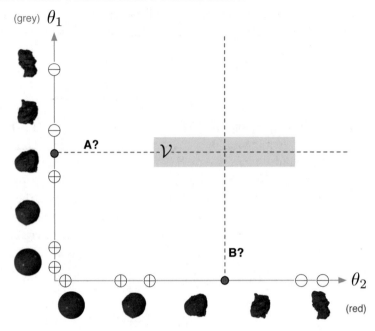

Figure 3.8: A problematic case for hypothesis space search. In this variant of the alien fruits example, there are two parameters: one for each color of fruit. This plot represents the 2D hypothesis space $\mathcal{H}$. Candidate queries A (grey) and B (red) both bisect the version space $\mathcal{V}$ (highlighted), and thus appear to be equally informative. However, the query that should most reduce generalization error depends on the distribution of fruit shapes.

query, because currently more red fruits are likely to be misclassified by θ_2 than grey fruits by θ_1 (due to the tightness of the labeled fruits providing bounds on them). However, grey fruits might be 10,000 times more common than red fruits, in which case A will probably yield better overall classification by making more use of a better estimate for θ_1. On the other hand, even if the fruit colors are highly skewed, the red fruits might be normally distributed with a mean near B, while the distribution of grey fruits might be bimodal with means far away from A. In this case, B is probably the informative query.

In pathological cases, both uncertainty sampling and QBC might exert a lot of effort querying in sparse, noisy, or irrelevant regions of the input space. This is especially true for query synthesis scenarios (see the discussion in Section 1.3), but it can also be problematic for the pool-based setting where the size of $\mathcal{U}$ is very large and full of statistical outliers. This is less of a problem in the stream-based selective sampling scenario, where data are sampled i.i.d. from the underlying distribution, and in fact most of the strong theoretical results in active learning (see Chapter 6) assume the stream-based setting. However, the active learning heuristics we have covered thus far are a bit myopic and can be prone to querying outliers. If our real desire is to develop an active learning system that reduces classification error, then perhaps we should consider active learning heuristics that try to optimize for that directly, which is the topic of the next chapter.

CHAPTER 4

Minimizing Expected Error and Variance

"The role of evidence is, in the main, to correct our mistakes, our prejudices, our tentative theories — that is, to play a part in the critical discussion, in the elimination of error."

— *Karl Raimund Popper*

4.1 EXPECTED ERROR REDUCTION

The task of an active learner is to identify the "best" instance to query. But the best in what sense? So far in this book we have looked at heuristics for selecting instances based on their uncertainty (Chapter 2) or their ability to reduce the hypothesis space (Chapter 3). What if we do not necessarily care about the model's certainty or the correctness of its hypotheses? Perhaps we really only care about most is how well it makes *predictions*, and neither of the previous frameworks directly optimize this notion of "best." With this in mind, let us shift attention away from what the learner thinks about instances *now* to what kinds of decisions it might make in the *future*. In particular, we want a learner to choose questions that, once it knows the answer, is most likely to reduce future error.

The problem is that the learner does not know what the answer to its question will be before asking it. Furthermore, do it does not know what its error will be even after it receives an answer and updates its hypotheses. Thus, we are forced to make a decision under uncertainty, and this is where decision theory comes in handy: we cannot reduce error as a known value, but we can try to minimize it as an *expected value*. The idea is that, when faced with several actions (each of which could result in different outcomes with different probabilities), one can identify all the possible outcomes, determine their values and probabilities, and compute a weighted sum to give an expected value for each action: $\mathbb{E}[V] = \sum_v P(v)v$. The "rational" decision should be to choose the action that results in the best expected value. In our case, this would be the lowest expected future error.

To compute the expected error, we need two probability distributions: (1) the probability of the oracle's label y in answer to query x; and (2) the probability that the learner will make an error on some other instance x' once the answer is known. The bad news is that neither of these are really known, but in both cases we can use the model's posterior distribution as a reasonable approximation. If we have a large unlabeled pool $\mathcal{U}$ available, the learner can attempt to minimize expected error over it, assuming that it is representative of the test distribution (using it as a sort of validation set).

To minimize the expected classification error (or 0/1-loss) over the unlabeled data $\mathcal{U}$, the decision-theoretic utility measure would look like this:

$$
\begin{aligned}
x_{ER}^* &= \underset{x}{\operatorname{argmin}} \; \mathbb{E}_{Y|\theta,x} \left[\sum_{x'\in\mathcal{U}} \mathbb{E}_{Y|\theta^+,x'}[y \neq \hat{y}] \right] \\
&= \underset{x}{\operatorname{argmin}} \; \sum_{y} P_\theta(y|x) \left[\sum_{x'\in\mathcal{U}} 1 - p_{\theta^+}(\hat{y}|x') \right],
\end{aligned}
\tag{4.1}
$$

where θ^+ refers to the a new model after it has been re-trained using a new labeled set $\mathcal{L} \cup \langle x, y\rangle$ — that is, after adding the candidate query x and the hypothetical oracle response y. The objective here is to reduce the expected total number of incorrect predictions (giving no credit for "near misses"). Another less stringent objective is to minimize the expected log-loss:

$$
\begin{aligned}
x_{LL}^* &= \underset{x}{\operatorname{argmin}} \; \mathbb{E}_{Y|\theta,x} \left[\sum_{x'\in\mathcal{U}} \mathbb{E}_{Y|\theta^+,x'}[-\log p_{\theta^+}(y|x')] \right] \\
&= \underset{x}{\operatorname{argmin}} \; \sum_{y} P_\theta(y|x) \left[\sum_{x'\in\mathcal{U}} -\sum_{y'} p_{\theta^+}(y'|x') \log p_{\theta^+}(y'|x') \right] \tag{4.2} \\
&= \underset{x}{\operatorname{argmin}} \; \sum_{y} P_\theta(y|x) \sum_{x'\in\mathcal{U}} H_{\theta^+}(Y|x'), \tag{4.3}
\end{aligned}
$$

which is equivalent to the expected total future output entropy (uncertainty) over $\mathcal{U}$.

Roy and McCallum (2001) proposed the expected error reduction framework for text classification using naive Bayes, and some of their results are presented in Figure 4.1. Here we see that sampling based on expected error reduction (log-loss in their case, Equation 4.2) produced significantly better learning curves in terms of accuracy than the other query strategies they tried: density-weighted query by committee (a variant of QBC which we will discuss in Chapter 5) and uncertainty sampling. Note that these active approaches are all still superior to random sampling ("passive" learning), but expected error reduction produces more accurate classifiers with less labeled data.

In most cases, unfortunately, expected error reduction is very computationally expensive. Not only does it require estimating the expected future error over $\mathcal{U}$ for each query, but a new model must be re-trained for every possible labeling of every possible query in the pool. This leads to a drastic increase in computational cost. For a classifier like naive Bayes, incremental re-training is fairly efficient since it only involves incrementing or decrementing a few counts. However, Roy and McCallum still had to resort to the bagging algorithm (Breiman, 1996) to ensure that the classifier's

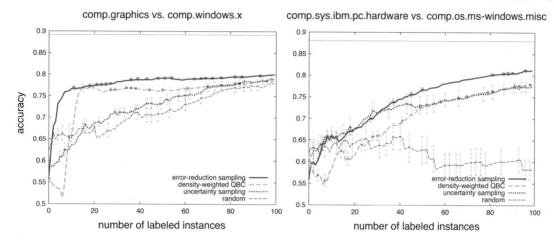

Figure 4.1: Learning curves showing that expected error reduction can outperform QBC and uncertainty sampling for two binary text classification tasks. *Source*: Adapted from Roy and McCallum (2001), with kind permission of the authors.

posterior distributions were good estimates[1]. Bagging helps to keep these estimates smooth and reliable, but adds to its computational overhead. For non-parametric model classes like Gaussian random fields or nearest-neighbor methods, the incremental training procedure is also efficient, making the approach fairly practical[2]. For a many other learning algorithms, however, this is not the case. For example, a binary logistic regression model would require $O(ULG)$ time simply to choose the next query, where U is the size of the unlabeled pool $\mathcal{U}$, L is the size of the current training set $\mathcal{L}$, and G is the number of gradient computations required by the by optimization procedure until convergence. A classification task with three or more labels using a maximum entropy model (Berger et al., 1996) would require $O(M^2 ULG)$ time complexity, where M is the number of class labels. For a sequence labeling task using conditional random fields (Lafferty et al., 2001), the complexity explodes to $O(TM^{T+2}ULG)$, where T is the length of an input sequence. Because of this, research papers employing expected error reduction have mostly only considered simple binary classification. Moreover, since the approach is often still impractical, researchers must resort to sub-sampling from the pool (Roy and McCallum, 2001) to reduce the U term in the above analyses, or use approximate training techniques (Guo and Greiner, 2007) to reduce the G term.

Nevertheless, there are some interesting variants of the approach. Zhu et al. (2003) combined this framework with semi-supervised learning (more on this in Chapter 5), resulting in a dramatic improvement over random or uncertainty sampling. Guo and Greiner (2007) employed an "opti-

---

[1] It is well known that, for problems where input features are not conditionally independent given the class label, a "double counting" effect can occur with naive Bayes. This violates the "naive" independence assumption, which can result in very sharp posteriors near to zero or one.

[2] The bottleneck in non-parametric models generally not re-training, but predictive inference.

mistic" variant that biases the expectation toward the most likely label for computational convenience, using uncertainty sampling as a fallback strategy when the oracle provides an unexpected labeling. This worked surprisingly well on several common data mining data sets from the UCI repository[3]. The estimated error reduction framework has the dual advantage of being near-optimal and not being dependent on the model class. All that is required is an appropriate objective function and a way to estimate posterior label probabilities. For example, strategies in this framework have been successfully used with a variety of models from naive Bayes and Gaussian random fields to logistic regression, and support vector machines. Furthermore, the general decision-theoretic approach can be employed not only to minimize error, as with 0/1-loss or log-loss, but to optimize any generic performance measure of interest, such as maximizing precision, recall, F_1-measure, or area under the ROC curve. As always, however, there are concerns of computational complexity, which may make these approaches infeasible for a truly interactive learning deployment.

4.2   VARIANCE REDUCTION

Minimizing an error function directly is costly, and in general it cannot be done in closed form. That is to say, the model must to be re-trained using hypothetical labelings in order to estimate the expected reduction in error (which can be computationally expensive). However, in some cases we may still be able to reduce generalization error *indirectly* by minimizing output variance, and this approach sometimes does have a closed-form solution. Consider a regression problem where the learning objective is to minimize standard error (i.e., squared-loss). We can take advantage of the following result from Geman et al. (1992), showing that a learner's expected error decomposes into:

$$
\begin{aligned}
\mathbb{E}\left[(\hat{y} - y)^2 | x\right] \;=\; & \mathbb{E}_{Y|x}\left[\,(y - \mathbb{E}_{Y|x}[y|x])^2\,\right] \\
& + \left(\,\mathbb{E}_{\mathcal{L}}[\hat{y}] - \mathbb{E}_{Y|x}[y|x]\,\right)^2 \\
& + \mathbb{E}_{\mathcal{L}}\left[\,(\hat{y} - \mathbb{E}_{\mathcal{L}}[\hat{y}])^2\,\right],
\end{aligned}
$$

where $\mathbb{E}_{Y|x}[\cdot]$ is an expectation over the conditional density $P(y|x)$, $\mathbb{E}_{\mathcal{L}}[\cdot]$ is an expectation over the labeled training set $\mathcal{L}$, and $\mathbb{E}[\cdot]$ is an expectation over both. Recall that $\hat{y}$ is shorthand for the model's prediction for a given instance x, while y indicates the true label for that instance.

The first term on the right-hand side of this equation is *noise*, i.e., the unreliability of the true label y given only x, which does not depend on the model or training data. Such noise may result from stochastic effects of the method used to obtain the labels, for example, or because the feature representation is inadequate. The second term is the *bias*, which represents the error due to the model class itself, e.g., if a linear model is used to learn a function that is non-linear. This component of the overall error is invariant given a fixed model class. The third term is the output *variance*, which is the remaining component of the learner's squared-loss with respect to the target

[3]http://archive.ics.uci.edu/ml/

function. Minimizing the variance, then, is guaranteed to minimize the future generalization error of the model (since the learner itself can do nothing about the noise or bias components).

Therefore, we can attempt to reduce error in the squared-loss sense by labeling instances that are expected to most reduce the model's output variance over the unlabeled instances $\mathcal{U}$:

$$x^*_{VR} = \operatorname*{argmin}_{x} \sum_{x' \in \mathcal{U}} \mathrm{Var}_{\theta^+}(Y|x'), \tag{4.4}$$

where θ^+ again denotes the model after it has been re-trained with $\mathcal{L} \cup \langle x, y \rangle$. The question is how to compute this value more efficiently than we could by directly minimizing the error through re-training, as we did in the previous section.

As it turns out, there is a rich history of such approaches in the statistics literature, often referred to as *optimal experimental design* (Chaloner and Verdinelli, 1995; Federov, 1972), and typically focused on regression tasks with a few input variables as predictors. However, the approaches are fairly general, and can be used to derive elegant closed-form utility functions for active learning. A key ingredient of these approaches is *Fisher information*, which is a way of measuring the amount of information that a random variable Y carries about a parameter θ upon which the likelihood function $P_\theta(Y)$ depends. The partial derivative of the logarithm of this function with respect to θ is called the *Fisher score*:

$$\nabla x = \frac{\partial}{\partial \theta} \log P_\theta(Y|x),$$

where ∇x denotes the score for an input instance x, upon which the output variable Y also depends. Note that the score does not depend on the actual label of x, only the distribution over Y under the parameters θ. Note also that, for multiple parameters, the score is a vector (or gradient). The Fisher information F is the variance of the score:

$$
\begin{aligned}
F &= \mathbb{E}_X\left[\left(\frac{\partial}{\partial \theta} \log P_\theta(Y|x)\right)^2\right] \\
&= \mathbb{E}_X\left[\frac{\partial^2}{\partial \theta^2} \log P_\theta(Y|x)\right] \\
&\propto \sum_x \nabla x \nabla x^\top.
\end{aligned}
$$

The Fisher score and information could also be written $\nabla_\theta x$ and $F(\theta)$, respectively, to make their relationship to model parameters more explicit, but I will use simplified notation here for brevity. Notice also that the Fisher information is not a function of a particular observation, as we have integrated over all instances in a particular input distribution.

Fisher information is important because its inverse sets a lower bound on the variance of the model's parameter estimates; this result is known as the Cramér-Rao inequality (Cover and Thomas, 2006). In other words, to minimize the variance over its parameter estimates, an active learner should select data that maximizes the Fisher information (or minimizes the inverse thereof). When there is only one parameter, the Fisher information is a scalar value and this strategy is straightforward. But for models of K parameters, Fisher information takes the form of a $K \times K$ covariance matrix, and deciding what exactly to optimize is less clear. In the literature, there are three main types of so-called optimal experimental designs:

- *D-optimality* minimizes the *determinant* of the inverse information matrix;

- *E-optimality* maximizes the minimum *eigenvalue* of the information matrix;

- *A-optimality* minimizes the trace of the inverse information matrix. This criterion results in minimizing the *average* variance of the parameter estimates.

D-optimality is related to minimizing the differential posterior entropy of the parameter estimates (Chaloner and Verdinelli, 1995), and results in the following utility function:

$$x_D^* = \underset{x}{\operatorname{argmin}} \ \mathbf{det}\left([F_{\mathcal{L}} + \nabla x \nabla x^\top]^{-1}\right),$$

where the additive property of Fisher information allows us to simply add the information content of x to that of all the previous training observations in $\mathcal{L}$. Note that this is a closed-form solution, and no actual model re-training is necessary to compute this measure of model variance. Since the determinant can be thought of as a measure of volume, the D-optimal design criterion can be thought of as the volume of the (noisy) version space, making it analogous to the query by committee algorithm (Chapter 3); it aims to select instances that reduce the amount of uncertainty in the parameter estimates. *E*-optimality does not seem to correspond to an obvious utility function, and is not often used in the machine learning literature, though there are some exceptions (e.g., Flaherty et al., 2006).

A-optimal designs are considerably more popular, and aim to reduce the *average* variance of parameter estimates by focusing on values along the diagonal of the information matrix. A common variant of *A*-optimal design is to minimize $\mathbf{tr}(A F_{\mathcal{L}}^{-1})$ — the trace of the inverse product of A and the information matrix — where A is a square, symmetric "reference" matrix. As a special case, consider a matrix of rank one, in particular: $A_x = \nabla x \nabla x^\top$. In this case, we have $\mathbf{tr}(A_x F_{\mathcal{L}}^{-1}) = \nabla x^\top F_{\mathcal{L}}^{-1} \nabla x$, which is the equation for computing the output variance for a single instance x in regression models (Schervish, 1995). This criterion is sometimes called *c*-optimality (where the vector $c = \nabla x$), which is an attempt to minimize the prediction variance for a single data instance. One way to do this is to simply query x itself, if we assume that the output variance will be zero once the true label is known. In this way, *c*-optimality can be viewed as a form of uncertainty sampling. For example, this is the strategy employed by the neural network in learning a toy 1D regression task in Figure 2.6.

Now recall that we are really interested in reducing variance across the input distribution (not merely a single data point), thus the A matrix should encode the whole instance space. Using

A-optimal design, we can derive an appropriate utility measure for active learning, called the *Fisher information ratio*:

$$
\begin{aligned}
x^*_{FIR} &= \underset{x}{\mathrm{argmin}} \sum_{x' \in \mathcal{U}} \mathrm{Var}_{\theta^+}(Y|x') \\
&= \underset{x}{\mathrm{argmin}} \sum_{x' \in \mathcal{U}} \mathbf{tr}(A_{x'}[F_{\mathcal{L}} + \nabla x \nabla x^{\top}]^{-1}) & (4.5) \\
&= \underset{x}{\mathrm{argmin}} \; \mathbf{tr}(F_{\mathcal{U}}[F_{\mathcal{L}} + \nabla x \nabla x^{\top}]^{-1}). & (4.6)
\end{aligned}
$$

We begin with the same objective as Equation 4.4, where $\mathrm{Var}_{\theta^+}(\cdot)$ denotes the variance of the model after it has be retrained with the query x and its putative label y. However, unlike in the previous section we now have a way of estimating this value in closed-form without explicit model re-training, by simply adding the information content of x to that of the current training set $\mathcal{L}$ (4.5). Furthermore, since matrix traces are also additive, the utility measure simplifies to the inner product of two matrices (4.6): the Fisher information $F_{\mathcal{U}} = \sum_{x' \in \mathcal{U}} A_{x'}$ encoded by the model's label predictions for $\mathcal{U}$, and the Fisher information for the label predictions for $\mathcal{L} \cup \{x\}$. This ratio can be interpreted as the model's output variance across the input distribution (represented by $\mathcal{U}$) that cannot yet be explained by the observations in $\mathcal{L} \cup \{x\}$. Querying the instance which minimizes this ratio is analogous to minimizing the future output variance once x has been labeled, thus indirectly reducing generalization error (with respect to $\mathcal{U}$) in the squared-loss sense. The advantage here over explicit error reduction, as we did at the beginning of this chapter, is that the model need not be retrained for each possible labeling of each candidate query. The information matrices give us an approximation of the updated output variance that simulates retraining.

MacKay (1992) derived such a utility measure to do active learning in regression tasks with neural networks, and Cohn (1994) applied it to a robot arm kinematics problem, predicting the absolute coordinates of a robot hand given the arm joint angles as inputs. The approach was actually a query synthesis scenario: by using numerical optimization, Cohn was able to find a set of joint angles that minimized variance among predictions across all legal joint angle combinations. The approach was computationally expensive and approximate, but Cohn et al. (1996) later showed that variance reduction active learning can be efficient and exact for statistical models like Gaussian mixtures and locally-weighted linear regression. For classification, Zhang and Oles (2000) and Schein and Ungar (2007) derived similar *A*-optimal heuristics for active learning with logistic regression. Hoi et al. (2006a) extended this idea to active text classification in the batch-mode setting (Section 4.3) in which a set of queries $\mathcal{Q}$ is selected all at once in an attempt to minimize the ratio between $F_{\mathcal{U}}$ and $F_{\mathcal{Q}}$. Settles and Craven (2008) also generalized the Fisher information ratio approach for structured prediction using conditional random fields.

There are still several practical drawbacks to variance-reduction methods, in terms of computational complexity. Estimating output variance requires inverting and multiplying $K \times K$ matrices, where K is the number of parameters in the model θ. Assuming standard implementations, these operations require $O(K^3)$ time[4]. This quickly becomes intractable for large K, which is a common occurrence in, say, natural language processing tasks (with hundreds of thousands or millions of parameters). Because these operations must be repeated for every instance in $\mathcal{U}$ being considered for querying, the computational complexity remains $O(UK^3)$ for selecting the next query. Paaß and Kindermann (1995) proposed a sampling approach based on Markov chains to reduce the U term in this analysis. To invert the information matrix and reduce the K^3 term, Hoi et al. (2006a) used principal component analysis to reduce the dimensionality of the parameter space. Once the information matrix has been inverted $F_{\mathcal{L}}^{-1}$, it can be efficiently updated using the Woodbury matrix identity (also known as the matrix inversion lemma) to incorporate information from the query instance $[F_{\mathcal{L}} + \nabla x \nabla x^{\top}]^{-1}$. Alternatively, Settles and Craven (2008) approximated the matrix with its diagonal vector, which can be inverted in only $O(K)$ time and results in substantial speedups. This is also more space efficient, since for very large K the square matrices may not even fit in memory. Empirically, these heuristics are still orders of magnitude slower than simple query strategies like uncertainty sampling. As a consequence, there are few experimental comparisons with other active learning methods, and the few that exist (Schein and Ungar, 2007; Settles and Craven, 2008) have reported mixed results.

4.3   BATCH QUERIES AND SUBMODULARITY

In most active learning research, queries are selected in *serial*, i.e., one at a time. However, sometimes the time required to induce a model is slow or expensive, as with large ensemble methods, or complex models for structured prediction tasks. Even if training can converge in a reasonable amount of time, active learning utility functions based on error and variance reduction techniques can sometimes be prohibitively expensive, as we have seen. It may also be that we have a finite budget of queries, such as deciding where to place a set of ten physical sensors in a room, and we must decide where to place them at the same time. In other cases, a distributed labeling environment may be available, e.g., multiple annotators working from different workstations at the same time on a computer network. In settings such as these, selecting queries in serial may be inefficient. By contrast, *batch-mode* active learning allows the learner to query instances in groups, which is arguably better suited to parallel labeling environments or models with slow training procedures.

The challenge in batch-mode active learning is how to properly assemble the optimal query set $\mathcal{Q}$. Myopically querying the "Q-best" queries according to some instance-level query strategy often does not work well, since it fails to consider the overlap in information content among the "best" instances. For example, if we query the two instances expected to most reduce output variance, they may both have high utility because they are virtually identical (and therefore redundant). Labeling one of these instances will be as informative as getting a label for the other, and we will waste effort by obtaining labels for them both. To address this, a few batch-mode active learning algorithms have been proposed. Brinker (2003) considered an approach for SVMs that explicitly incorporates diversity among instances in the batch. Xu et al. (2007) proposed a similar approach for SVM active

---

[4]Certain optimizations can reduce the exponent, but known matrix algorithms are still more than quadratic.

learning, which also incorporates a density measure (Chapter 5). Specifically, they query the centroids of clusters of instances that lie closest to the decision boundary. Alternatively, Guo and Schuurmans (2008) treated batch construction for logistic regression as an optimization problem, and attempt to construct the most informative batch directly through a gradient search. For the most part, these approaches show improvements over random batch sampling, which in turn is generally better than simple "Q-best" batch construction.

As it turns out, though, active learning heuristics based on variance reduction lend themselves quite naturally to the batch setting, sometimes with performance guarantees. This can be done by exploiting the properties of *submodular functions* (Nemhauser et al., 1978). Submodularity is a property of set functions that intuitively formalizes the idea of "diminishing returns." That is, adding some instance x to the set $\mathcal{A}$ provides more gain in terms of the utility function than adding x to a larger set $\mathcal{A}'$, where $\mathcal{A} \subseteq \mathcal{A}'$. Informally, since $\mathcal{A}'$ is a superset of $\mathcal{A}$ and already contains more data, adding x will not help as much. More formally, a set function s is submodular if it satisfies the property:

$$s(\mathcal{A} \cup \{x\}) - s(\mathcal{A}) \ \geq \ s(\mathcal{A}' \cup \{x\}) - s(\mathcal{A}'),$$

or, equivalently:

$$s(\mathcal{A}) + s(\mathcal{B}) \ \geq \ s(\mathcal{A} \cup \mathcal{B}) + s(\mathcal{A} \cap \mathcal{B}),$$

for any two sets $\mathcal{A}$ and $\mathcal{B}$. The key advantage of submodularity is that, for monotonically non-decreasing submodular functions where for the empty set $s(\emptyset) = 0$, a greedy algorithm for selecting N instances guarantees a performance of $(1 - 1/e) \times s(\mathcal{A}_N^*)$, where $s(\mathcal{A}_N^*)$ is the value of the optimal set of size N. In other words, using a greedy algorithm to incrementally pick items to optimize a submodular function will give us a lower-bound performance guarantee of around 63% of optimal. In practice, greedy algorithms for submodular criteria can be within 90% of optimal (Krause, 2008).

A variance reduction heuristic can be recast as a monotonically non-decreasing function by measuring the total difference between the model's output variance before querying the set Q and the expected variance afterward:

$$s(Q) \ = \ \sum_{x \in \mathcal{U}} \text{Var}_\theta(Y|x) - \text{Var}_{\theta + Q}(Y|x)$$

$$= \ \mathbf{tr}(F_{\mathcal{U}} F_{\mathcal{L}}^{-1}) - \mathbf{tr}(F_{\mathcal{U}}[F_{\mathcal{L}} + F_Q]^{-1}).$$

For $Q = \emptyset$, this utility measure is clearly zero, and we can greedily select the instance x that maximizes it, add x to the set Q, and repeat until the query budget is spent. Under certain conditions with certain learning algorithms, this approach can be shown to be submodular. Some examples are Gaussian processes (Guestrin et al., 2005), logistic regression (Hoi et al., 2006b), and linear regression (Das and Kempe, 2008). In fact, for linear regression the Fisher score and Fisher information matrices depend only on the input vector x and not on the parameter coefficients. Thus, all queries within a given budget Q can be selected at the outset before *any* data has been labeled (MacKay, 1992)! In settings where there is a fixed budget for gathering data — or where it is more practical to gather the training data in batches — submodular utility measures guarantee near-optimal results with significantly less computational effort[5].

4.4 DISCUSSION

In this chapter, we touched on several principled active learning strategies that aim to directly minimize the expected error or output variance on instances in the unlabeled distribution. These kinds of approaches are intuitively appealing, and have enjoyed some empirical success in terms of producing more accurate learners with fewer labeled instances than uncertainty or hypothesis-based approaches. Methods that aim to reduce classification error by minimizing 0/1-loss or log-loss (Section 4.1) cannot be solved in closed form generally, and require instead that models be re-trained to account for all possible labelings of all possible queries. In contrast, methods that aim to reduce output variance in the squared-loss sense (Section 4.2) have a long history in the statistics literature, can be computed in closed form, and apply equally well to classification and regression problems. They can also be generalized in a rather straightforward way to perform active learning in batches (sometimes with efficiency guarantees based on properties of submodularity).

Still, there are two major drawbacks to these approaches. The first is that they are not as general as uncertainty sampling or disagreement-based methods, and can only be easily applied to hypothesis classes of a certain form. In particular, the variance reduction heuristics are only natural for linear, non-linear, and logistic regression models (which is not surprising, since they were designed by statisticians with these models in mind). How to use them with decision trees, nearest-neighbor methods, and a variety of other standard machine learning and data mining techniques is less clear. The second drawback is their computational complexity due to the inversion and multiplication of the Fisher information matrix. If the number of parameters or output variables is relatively small, then these methods can still be relatively fast and efficient. However, if these properties of the problem are very large then both the space and time complexities of the algorithms involved might be too high to be useable in practice.

---

[5]Many set optimization problems such as this are NP-hard, resulting in computations that unfortunately scale exponentially with the size of the input. Therefore, greedy approaches are usually more efficient.

CHAPTER 5

Exploiting Structure in Data

"A rare pattern can contain an enormous amount of information, provided that it is closely linked to the structure of the problem space."

— *Allen Newell and Herbert A. Simon*

5.1 DENSITY-WEIGHTED METHODS

Strategies like uncertainty sampling (Chapter 2) and query by committee (Chapter 3), at least employed in a pool-based setting, are somewhat myopic: they only measure the information content of a single data instance. As a result, these approaches run the risk of selecting outliers and other poor choices for queries. Figure 5.1 illustrates this problem for a binary linear classifier using uncertainty sampling. The least certain instance lies on the classification boundary, but is not very representative of the input distribution. So knowing its label is unlikely to improve the classifier's accuracy on the data as a whole. Query by committee (QBC) can exhibit similar behavior. In fact, it may be that these instances are uncertain or controversial precisely because they are outliers.

Figure 5.1: An illustration of when uncertainty sampling can be a poor strategy. Shaded polygons represent labeled instances in $\mathcal{L}$, and circles represent unlabeled instances in $\mathcal{U}$. Since A is on the decision boundary, it would be queried as the most uncertain. However, B would probably provide more information about the input distribution as a whole.

In contrast, error and variance reduction approaches (Chapter 4) attempt to reduce the expected future error over all data instances, implicitly taking the input distribution into account. By utilizing the unlabeled pool $\mathcal{U}$ when estimating future errors or output variances, these heuristics can be less sensitive to outliers and noisy inputs than their simpler relatives like uncertainty sampling and QBC. However, these advantages come at a significant computational expense, and in some cases the costs may not be worth the potential gains. In this chapter, we look at ways of overcoming these problems by explicitly considering the structure of the data while selecting queries.

Density-weighted heuristics for active learning aim to do exactly this. The basic intuition is that informative instances should not only be those with high information content, but also those which are representative of the underlying distribution in some sense (e.g., inhabit dense regions of the input space). The generic *information density* heuristic captures the main ideas of this approach:

$$x_{ID}^* = \operatorname*{argmax}_{x} \; \phi_A(x) \times \left(\frac{1}{U} \sum_{x' \in \mathcal{U}} sim(x, x') \right)^{\beta} . \tag{5.1}$$

Here, U is the size of the unlabeled pool and $\phi_A(\cdot)$ represents the utility of x according to some "base" query strategy A, such as an uncertainty sampling or QBC approach. For example, one could use prediction entropy as the base utility measure: $\phi_H(x) = H_\theta(Y|x)$. The second term weights the informativeness of x by its average similarity to all other instances in the input distribution — for which $\mathcal{U}$ is assumed to be representative — subject to a parameter β that controls the relative importance of the density term. For example, one might use cosine similarity, Euclidean distance, Pearson's correlation coefficient, Spearman's rank correlation, or any other problem-appropriate measure of similarity among instances.

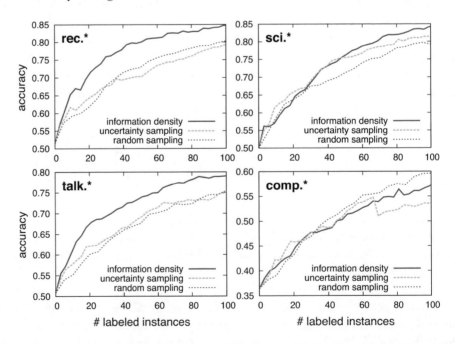

Figure 5.2: Learning curves showing that, by explicitly weighting queries by their representativeness among the input instances, information density can yield better results than the base uncertainty sampling heuristic by itself.

Figure 5.2 shows learning curves for four subsets of the famous 20 Newsgroups text classification corpus (Lang, 1995). The experiments compare information density (using entropy as the base utility measure, cosine similarity, and $\beta = 1$) to simple entropy-based uncertainty sampling, and random sampling. All results are averaged across ten folds using cross-validation. Even in cases where uncertainty sampling is worse than random sampling, information density can yield comparable or superior results. Such density-weighted approaches have been proposed and employed in a variety of machine learning applications (Fujii et al., 1998; McCallum and Nigam, 1998; Settles and Craven, 2008; Xu et al., 2007). In general, the published results seem to indicate that these approaches are superior to simpler methods that do not consider density or representativeness in their calculations. Although there are some exceptions, such as the comp.* experiments in Figure 5.2, as well as in Figure 4.1 (where Roy and McCallum (2001) used a density-weighted QBC approach as a baseline for their error-reduction utility measure). However, a big advantage of density-weighting over error and variance reduction is this: if the density terms are pre-computed and cached for efficient lookup, the time required to select the next query is essentially no different than the base informativeness measure, e.g., uncertainty sampling (Settles and Craven, 2008). This is advantageous for conducting active learning interactively with oracles in real-time.

5.2   CLUSTER-BASED ACTIVE LEARNING

Exploiting the input distribution also brings to mind interesting unsupervised learning techniques, like clustering, and how they might be incorporated into active learning (see Figure 5.3). For example, variants of density-weighting might first cluster the pool $\mathcal{U}$ and compute average similarities of each instance to all the other instances in the same cluster. Depending on the data and the clustering algorithm used, this might in fact be faster than a full $O(N^2)$ comparison for large data sets. Another idea is to pre-cluster the data and begin by querying cluster centroids to comprise the initial $\mathcal{L}$, in effect "warm starting" active learning instead of beginning with a random sample (Kang et al., 2004). One could also cluster the most informative instances after each iteration and query the most representative instances of those clusters (Nguyen and Smeulders, 2004).

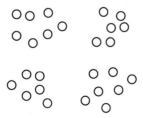

Figure 5.3: A motivating example for cluster-based active learning. If the input distribution looks like this, perhaps we only need four labels?

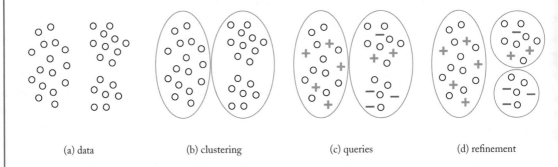

| (a) data | (b) clustering | (c) queries | (d) refinement |

Figure 5.4: The basic stages of active learning by hierarchical sampling.

However, naive cluster-based approaches have their limitations. First, there may not be an obvious clustering of $\mathcal{U}$, or a good similarity measure for clustering the data is unknown. Second, good clusterings may exist at various levels of granularity (should there be four clusters, or forty?). Third, the clusters may not correspond to the hidden class labels after all. To take the "best of both worlds," it would be nice to have an active learning algorithm that makes no hard assumptions about the relationship between input and label distributions, but is still able to take advantage of data clusterings when they happen to be informative. In fact, let us consider an active learning algorithm that selects queries based solely on the cluster structure and labels received so far, and is therefore agnostic about the type of classifier (i.e., model or hypothesis class) we intend to use. The basic idea is illustrated in Figure 5.4. First we (a) take the data and (b) find an initial coarse clustering. Then we (c) query instances from these clusters and (d) iteratively refine the clusterings so that they become more pure, and focus our querying attention on the impure clusters.

This approach is called *hierarchical sampling* (Dasgupta and Hsu, 2008), and the full algorithm is shown in Figure 5.5. The input is a hierarchical clustering $\mathbf{T}$ of the unlabeled data, where the subtree rooted at node $v \in \mathbf{T}$ is denoted $\mathbf{T}_v$. The algorithm maintains a pruning $\mathcal{P} \subset \mathbf{T}$, or a subset of cluster nodes that are disjoint and contain all the leaves (instances) — in other words, $\mathcal{P}$ consists of the topmost few layers of the hierarchical clustering. Each node $u \in \mathcal{P}$ is assigned its majority labeling $L(v)$ so that at any time, if the algorithm were interrupted, all the instances in $\mathbf{T}_v$ would be given this label. For example, the bottom-right cluster in Figure 5.4(d) would be assigned the negative label, whereas the other two would be assigned the positive label. These labelings are then used to construct a training set $\mathcal{L}$ for supervised learning algorithms. Clearly, this results in label noise if there are instances in $\mathbf{T}_v$ whose true labelings are not $L(v)$, as is the case in the upper-right-hand cluster in Figure 5.4(d). Therefore, we also want to maintain a bound $\check{p}_{v,y}$ on the lowest estimated probability of label y at node v, which could be estimated from empirical label frequencies, and used to estimate an upper-bound of the error at node v. This can be computed in a linear-time bottom-up pass through the tree and used to refine the pruning $\mathcal{P}$, for example, by selecting the

1: hierarchical clustering $\mathbf{T} = \mathbf{cluster}(\mathcal{U})$
2: pruning $\mathcal{P} = \{\text{root}(\mathbf{T})\}$
3: **for** $t = 1, 2, \ldots$ **do**
4:    cluster node $v = \mathbf{select}(\mathcal{P})$
5:    pick a random instance x from subtree $\mathbf{T}_v$ and query its label
6:    update counts for all cluster nodes u on a path from $x \to v$
7:    choose the best pruning $\mathcal{P}'_v$ and labeling L'_v for $\mathbf{T}_v$
8:    $\mathcal{P} = (\mathcal{P} - \{v\}) \cup \mathcal{P}'_v$
9:    $L(u) = L'_v(u)$ for all $u \in \mathcal{P}'_v$
10: **end for**
11: **for all** $v \in \mathcal{P}$ **do**
12:    add $\langle x, L(v) \rangle$ to labeled set $\mathcal{L}$ for all $x \in \mathbf{T}_v$
13: **end for**
14: return the labeled set $\mathcal{L}$ for training

Figure 5.5: The hierarchical sampling algorithm. For each node v in the clustering, the algorithm maintains the majority label $L(v)$, empirical label frequencies $\tilde{p}_{v,y}$, and the lower bound $\breve{p}_{v,y}$ on all label probabilities at node $\mathbf{T}_v$.

upper-right-hand cluster in Figure 5.4(d) with higher probability than the other two clusters (which are much more pure).

A key difference between this and most of the active learning algorithms in this book is that the learner queries *clusters*, and obtains labels for randomly drawn instances within a cluster. Regardless of how we choose to implement the **cluster**($\cdot$) and **select**($\cdot$) procedures (lines 1 and 4, respectively), the algorithm remains statistically sound: it maintains valid estimates for the error induced by its current pruning, because the labeled instances are drawn at random from the clusters. Therefore, we have a lot of freedom in how to perform these operations. Consider these two options for the **select**($\cdot$) procedure:

- select $v \in \mathcal{P}$ with probability $\propto |\mathbf{T}_v|$ — this is similar to random sampling;

- select $v \in \mathcal{P}$ with probability $\propto |\mathbf{T}_v|(1 - \breve{p}_{v,L(v)})$ — this is an active learning rule that focuses attention on clusters which appear to be impure or erroneous.

If the cluster structure is pathologically not correlated with the label structure — i.e., all the clusters are fairly impure — then the behavior of the algorithm will essentially resemble the first option above, and degrade gracefully to random sampling. However, if the clusterings do have a relationship with the hidden label distribution, we can take advantage of it by sampling more often from the impure clusters with the second option (lines 4–6), and refining the clusters accordingly (lines 7–9). If the clustering contains a pruning whose clusters are ϵ-pure in their labels, then the hierarchical

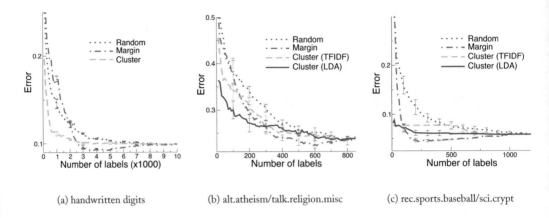

(a) handwritten digits   (b) alt.atheism/talk.religion.misc   (c) rec.sports.baseball/sci.crypt

Figure 5.6: Learning curves for cluster-based hierarchical sampling vs. margin (uncertainty) and random sampling baselines on (a) handwritten digit classification and (b–c) text classification. *Source*: Adapted from Dasgupta and Hsu (2008), with kind permission of the authors.

sampling algorithm can find a labeling that proportionally pure with $O(|\mathcal{P}|d(\mathcal{P})/\epsilon)$ labels, where $d(\mathcal{P})$ is the maximum depth of a node in the pruning $\mathcal{P}$. More details on these bounds are discussed in Dasgupta and Hsu (2008).

Figure 5.6 shows learning curves that compare hierarchical sampling to margin-based uncertainty sampling and random sampling (Dasgupta and Hsu, 2008). In these experiments **T** is constructed using an agglomerative bottom-up clustering algorithm with an average-linkage distance measure (Ward, 1963), and logistic regression is the final supervised learning algorithm. Figure 5.6(a) plots the classification error for the ten-label problem of classifying images in the MNIST handwritten digits data set[1], using Euclidean distance as the similarity measure for clustering. Figures 5.6(b–c) are on binary subsets of the 20 Newsgroups text classification corpus[2], using different document preprocessing techniques for clustering, namely TFIDF (Manning and Schütze, 1999) and latent Dirichlet allocation (Blei et al., 2003), a topic modeling approach. Both active learning approaches produce better learning curves than random sampling. The cluster-based hierarchical sampling manages to produce steeper learning curves early on than margin-based uncertainty sampling in all cases, but uncertainty sampling manages to achieve lower error rates later on in the process. This may be a case of serendipitous sampling bias for these particular data sets. However, hierarchical sampling avoids sampling bias by concentrating on converging to the same results as if it had all of the correct training labels.

---

[1] http://yann.lecun.com/exdb/mnist/
[2] http://people.csail.mit.edu/jrennie/20Newsgroups/

Table 5.1: Conceptually, several active learning algorithms have semi-supervised learning counterparts.

Active Learning	Semi-Supervised Learning
uncertainty sampling	self-training
query by committee	co-training
expected error reduction	entropy regularization

5.3   ACTIVE + SEMI-SUPERVISED LEARNING

Semi-supervised learning (Zhu and Goldberg, 2009), like active learning, aims to improve upon supervised machine learning methods by making the most of the vast amounts of unlabeled data that may be available. However, they take essentially complementary approaches. Whereas active learning aims to minimize labeling effort by posing queries of the most informative instances, semi-supervised learning aims to let the model "teach itself" by extrapolating what it thinks it has learned onto unlabeled instances. As such, semi-supervised methods try to exploit latent structure in the data to improve the quality of the learned model. As a result, there are a few conceptual overlaps between semi-supervised and active learning, as summarized in Table 5.1.

A very basic semi-supervised technique is *self-training* (Yarowsky, 1995), in which the learner is first trained with a small amount of labeled data, and then classifies the unlabeled data to re-train itself. This is typically done in an iterative fashion, where the *most* confident unlabeled instances, together with their predicted labels, are added to the training set $\mathcal{L}$, and the process repeats. The assumption is that the learner can trust its most confident predictions. A complementary technique in active learning is uncertainty sampling, where the instances about which the model is *least* confident are selected for querying, assuming that the learner needs the most guidance on these instances.

Alternatively, *co-training* (Blum and Mitchell, 1998) and *multi-view learning* (de Sa, 1994) use ensembles for semi-supervised learning. Initially, separate models are trained with the labeled data (usually using separate, conditionally independent feature sets), which then classify the unlabeled data, and "teach" the other models with a few unlabeled instances (using predicted labels) about which they are most confident. For example, consider the task of classifying web pages: one feature set might be the words contained on the page itself, while another feature set might consist of the words contained in hyperlinks to that page. Forcing these different views of the data to agree not only on the labeled data, but the unlabeled data as well — in this case, large numbers of web pages — helps to reduce the size of the version space. Query by committee is an active learning complement here, as the committee is meant to approximate different parts of the version space, and is used to query the unlabeled instances about which these competing hypotheses do *not* agree.

Finally, *entropy regularization* (Grandvalet and Bengio, 2005) is a semi-supervised learning approach based on the intuition that the best model of the data is the one that can make the most confident predictions on the unlabeled data. For example, consider training a logistic regression classifier by maximizing the following likelihood function ℓ:

$$\ell_\theta(\mathcal{L}, \mathcal{U}) = \sum_{\langle x, y \rangle \in \mathcal{L}} \log P_\theta(y|x) \; - \sum_k \frac{\theta_k^2}{2\sigma^2} \; - \sum_{x' \in \mathcal{U}} H_\theta(Y|x'). \qquad (5.2)$$

The first summand is the conditional log likelihood of the labeled data, while the second is the common L_2 regularization term to prevent over-fitting. These two terms comprise the standard objective function for supervised training. The third term aims to reduce the model's output entropy (e.g., the expected log-loss) among the unlabeled data. In other words, this entropy term penalizes parameter settings with the highest risk of making mistakes on the unlabeled data. In active learning, expected error reduction using log-loss (Equation 4.2) aims to select instances that yield the analogous result.

We can see through these examples that many popular active and semi-supervised learning algorithms try to attack the same problem — making the most of unlabeled data — from opposite directions. While semi-supervised methods exploit what the learner thinks it knows about the unlabeled data, active methods attempt to explore the unknown aspects[3]. It is therefore natural to think about combining the two approaches, which is often easily done. McCallum and Nigam (1998) combined density-weighted QBC for naive Bayes with the expectation-maximization (EM) algorithm, which can be seen as a kind of "soft" self-training. Muslea et al. (2002) combined multi-view algorithms for active and semi-supervised learning into a unified approach, also using naive Bayes as the underlying learning algorithm. Zhu et al. (2003) combined an error-reduction query selection strategy with a graph-based semi-supervised learning approach which also aims to minimize uncertainty (similar in spirit to entropy regularization), applied to text classification and handwritten digit recognition. Tomanek and Hahn (2009) combined self-training with uncertainty sampling to train linear-chain conditional random fields for information extraction, while Tür et al. (2005) combined active and semi-supervised learning for speech recognition. Other examples of active semi-supervised learning systems are described by Zhou et al. (2004) and Yu et al. (2006).

5.4   DISCUSSION

In this chapter, we considered ways in which the input distribution — whether it is known or can be modeled from a large pool of unlabeled data — can be explicitly taken advantage of in active learning. One simple approach is to combine a simple utility measure (like uncertainty or QBC) with a "representativeness" measure like input density. This can help the active learner avoid queries that are statistical outliers and otherwise atypical or spurious instances that have little to do with the problem. Another approach is to exploit cluster structure in the data, and either query clusters or cluster centroids to obtain labels that are both representative and diverse. Finally, active learning and semi-supervised learning both traffic in making the most of unlabeled data, but take largely complementary approaches. It therefore seems natural to combine them so that the learner can exploit both what it does and does not know about the problem.

[3]Some might argue that active methods are also "exploiting" what is known by querying what is *not* known, rather than "exploring" the data at random.

CHAPTER 6

Theory

"Theory is the first term in the Taylor series expansion of practice."

— *Thomas M. Cover*

6.1 A UNIFIED VIEW

So far, we have looked at a variety of heuristics by which an active learner might select instances to query, each motivated by slightly different intuitions or assumptions about the problem. To put these various query frameworks into perspective, let us consider a single, well-motivated utility measure that aims to maximize the amount of information we gain from that query. As it turns out, all of the general query frameworks we have looked at contain a popular utility function that can be viewed as an approximation to this measure under certain conditions.

If we assume that the unlabeled pool $\mathcal{U}$ is representative of the input distribution $\mathcal{D}_X$, and that the optimal sequence of queries can be obtained by greedily selecting the queries that maximize the information gain at each iteration, then our objective is:

$$x^* = \underset{x}{\text{argmax}} \sum_{x' \in \mathcal{U}} \Big(H_\theta(Y|x') - H_{\theta^+}(Y|x') \Big),$$

where θ^+ denotes a new model after the query and its associated label have been added to $\mathcal{L}$ and used to re-train the model. This measure represents the total gain in information (i.e., the change in entropy or uncertainty) over all the instances. Unfortunately, as we have already seen, the true label y is not known for the query x, so we again resort to the current model's predictions and compute the *expected* information gain:

$$x^* = \underset{x}{\text{argmax}} \sum_{x' \in \mathcal{U}} \Big(H_\theta(Y|x') - \mathbb{E}_{Y|\theta,x}\big[H_{\theta^+}(Y|x')\big] \Big). \tag{6.1}$$

From this utility measure, we can derive a simple uncertainty sampling heuristic:

$$x^* = \underset{x}{\text{argmax}} \sum_{x' \in \mathcal{U}} \Big(H_\theta(Y|x') - \mathbb{E}_{Y|\theta,x}\big[H_{\theta^+}(Y|x')\big] \Big)$$

$$\approx \underset{x}{\text{argmax}} \ H_\theta(Y|x) - \mathbb{E}_{Y|\theta,x}\big[H_{\theta^+}(Y|x)\big] \tag{6.2}$$

$$\approx \underset{x}{\text{argmax}} \ H_\theta(Y|x). \tag{6.3}$$

Hence, the *entropy*-based uncertainty sampling strategy from Equation 2.3 can be seen as an approximation of our idealized utility measure, under two (grossly over-simplifying) assumptions. The first approximation (6.2) stems from the idea that all the instances in $\mathcal{U}$ have equal impact on each other, so the information gain of query x will be proportional to any other instance x'. The second approximation (6.3) follows from intuition that, if the model is re-trained with the oracle's true label y for query x, then the label is known, it can be accurately predicted, and its entropy will be zero. Similarly, if our goal was to maximize the gain in 0/1-loss (as opposed to entropy here), these same approximations would yield the *least confident* uncertainty sampling strategy from Equation 2.1.

Query by committee (QBC) heuristics such as the *vote entropy* variants in Equations 3.1 and 3.2 have a very similar interpretation. Namely, it makes the assumptions that query x is representative of $\mathcal{U}$ and that the expected future entropy of that query instance, once labeled, is zero. The main difference is that QBC methods replace the point estimate θ with a distribution over several hypotheses $\theta \in \mathcal{C}$, approximating the version space with a committee. This Bayesian flavor helps to mitigate problems caused by our first assumption in this chapter (6.1): that prediction under the current single model θ yields a good estimate of the true label distribution.

The relationship of our ideal utility measure to the error-reduction framework in Chapter 4 is a bit more obvious:

$$
\begin{aligned}
x^* &= \operatorname*{argmax}_{x} \sum_{x' \in \mathcal{U}} \left(H_\theta(Y|x') - \mathbb{E}_{Y|\theta,x}\big[H_{\theta+}(Y|x')\big] \right) \\
&= \operatorname*{argmax}_{x} \sum_{x' \in \mathcal{U}} H_\theta(Y|x') - \sum_{x' \in \mathcal{U}} \mathbb{E}_{Y|\theta,x}\big[H_{\theta+}(Y|x')\big] \\
&= \operatorname*{argmin}_{x} \sum_{x' \in \mathcal{U}} \mathbb{E}_{Y|\theta,x}\big[H_{\theta+}(Y|x')\big]. &(6.4)
\end{aligned}
$$

Since the first entropy term $H_\theta(Y|x')$ depends only on the current state of the model θ, is it held constant for all possible queries and can be dropped from the equation. Thus, maximizing the negative expected future entropy is the same as minimizing the expected log-loss, making this equivalent to the expected reduction of log-loss utility measure from Equation 4.2. Since the entropy of a random variable is a monotonic function of its variance, variance-reduction utility measures such as Fisher information ratio (4.6) are also approximations to this ideal objective, in cases where the proper assumptions hold. If we wish to maximize the gain in 0/1-loss instead, the result is Equation 4.1.

Finally, let us consider how density-weighted methods from Chapter 5 can be derived from this ideal utility. The approximation is similar to that of uncertainty sampling (or QBC), but attenuates the overly strong assumption that a single instance x is representative of the input distribution. The expression inside the summand of Equation 6.1 represents the expected reduction in entropy for an instance $x' \in \mathcal{U}$ if x is the selected as the query. Let us hold to the assumption that the expected future entropy of x will be zero (since the oracle provides the label). However, let us make a less stringent assumption about the rest of the input distribution: the reduction in entropy for all other x' is proportional to its similarity with x. By this notion, we can make a simple substitution in the parenthetical:

$$x^* = \underset{x}{\operatorname{argmax}} \sum_{x' \in \mathcal{U}} \left(H_\theta(Y|x') - \mathbb{E}_{Y|\theta,x}\left[H_{\theta^+}(Y|x') \right] \right)$$

$$\approx \underset{x}{\operatorname{argmax}} \sum_{x' \in \mathcal{U}} \left(\operatorname{sim}(x, x') \times H_\theta(Y|x) \right). \tag{6.5}$$

With some trivial rearranging of terms, this is precisely the information density measure from Equation 5.1, if we use entropy as the base utility measure ϕ_A, set $\beta = 1$, and ignore the constant $\frac{1}{U}$. Hence, information density is can also be viewed as a somewhat scalable, but less harsh approximation to our idealized utility measure.

6.2   A PAC BOUND FOR ACTIVE LEARNING

It would be nice to have a bound on the number of queries required train a good classifier, as well as a guarantee that this number is smaller than the number of labeled instances required in the passive supervised setting. For example, at the beginning of this book we showed that for a 1D thresholding task (e.g., the alien fruits example), a binary search active learning algorithm guarantees an exponential reduction in the number of instances that need to be labeled. In this section, we will delve into a bit more detail and see if we can obtain analogous results for more general active learning problem settings.

First, a review of the *probably approximately correct* (PAC) learning setting (Valiant, 1984), a formal framework for mathematical analysis of machine learning. The goal is to learn, with high probability $1 - \delta$ (the *probably* part), a function that will have low future generalization error ϵ (the *approximately correct* part) on new data instances from the same input distribution. We will not cover all of the details of PAC learning here; see Mitchell (1997) or Vapnik (1998) for more thorough introductions. We are interested in finding an upper bound on the number of labeled instances that would be required to train a probably approximately correct classifier in this sense. Let L be the *label complexity*, defined as the smallest integer t such that for all $t' \geq t$,

$$P\left(\operatorname{err}(h_t) \leq \epsilon \right) \geq 1 - \delta,$$

where h_t is a hypothesis (classifier) output by the learning algorithm after t queries, and $\operatorname{err}(\cdot)$ is the error rate (expected 0/1-loss) of a classifier. In other words, the label complexity L is the number of labeled instances required to train a classifier that satisfies the desired values of ϵ and δ. In passive supervised learning, such label complexity bounds have been fairly well-studied.

Theorem 6.1   Let $\mathcal{D}_{XY}$ denote a joint instance-label distribution, and assume that labeled instances in the training set $\mathcal{L} \backsim \mathcal{D}_{XY}$ are i.i.d. samples. If $\mathcal{D}_{XY}$ is perfectly separable, i.e., there exists a hypothesis h^* such that $\text{err}(h^*) = 0$, then the label complexity L_{PASS} of a passive learning algorithm is (Vapnik, 1998):

$$L_{\text{PASS}} \leq O\left(\frac{1}{\epsilon}\left(d\log\frac{1}{\epsilon} + \log\frac{1}{\delta}\right)\right)$$

$$= \tilde{O}\left(\frac{d}{\epsilon}\right).$$

The $\tilde{O}$ notation simplifies things by ignoring terms that are logarithmic or dependent on δ (which is typically modest in supervised learning bounds). Here also d denotes the *VC dimension* (Vapnik and Chervonenkis, 1971), a parameter which is useful in characterizing the complexity of a hypothesis class $\mathcal{H}$. In brief, d represents the largest number of distinct instances that can be completely discriminated (or "shattered") by a hypothesis $h \in \mathcal{H}$. For example, linear thresholds on the 1D number line have VC dimension $d = 1$. Plugging this value into Theorem 6.1 shows why, in Chapter 1, we needed to test $\tilde{O}(1/\epsilon)$ fruits in order to passively train an accurate classifier for the alien fruits example. Other canonical VC dimensions include intervals on the 1D number line ($d = 2$), axis-parallel rectangles in K dimensions ($d = 2K$), and linear separators ($d = K + 1$).

Now let us consider the query by disagreement algorithm from Section 3.3 (Cohn et al., 1994). Recall that QBD is a selective sampling algorithm, meaning we draw instances in a stream from the marginal distribution $\mathcal{D}$ over the instance space (dropping $\mathcal{D}_X$'s explicit dependence on X for subsequent brevity), but the labels are hidden. It is the task of the learner to request labels for select instances according to the algorithm in Figure 3.3, which maintains all hypotheses in the version space and queries instances about which there is any disagreement. In order to proceed, there are a few formal definitions that will prove helpful in our analysis. First, the data distribution provides a natural measure of the "difference" $\Delta(h_1, h_2)$ between any two hypotheses, defined as:

$$\Delta(h_1, h_2) = P_{\mathcal{D}}(h_1(x) \neq h_2(x)).$$

Intuitively, Δ encodes the proportion of instances about which the two hypotheses disagree. The r-ball centered at some hypothesis h^* is defined as:

$$B(h^*, r) = \{h \in \mathcal{H} \mid \Delta(h^*, h) \leq r\},$$

that is, the set of all hypotheses for which the probability mass under the region of disagreement with h^* is within radius r. For a given version space $\mathcal{V}$ and instance space $\mathcal{X}$, the *region of disagreement* is defined as:

$$\text{DIS}(\mathcal{V}) = \{x \in \mathcal{X} \mid \exists\, h_1, h_2 \in \mathcal{V} : h_1(x) \neq h_2(x)\}.$$

This is the set of instances about which there is some disagreement among hypotheses in the version space. These three notions are visually depicted in Figure 6.1.

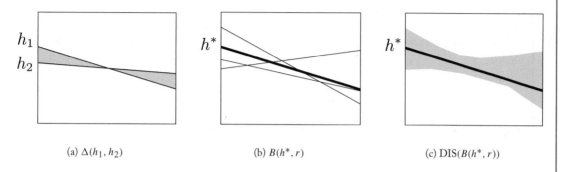

(a) $\Delta(h_1, h_2)$ (b) $B(h^*, r)$ (c) DIS$(B(h^*, r))$

Figure 6.1: Suppose the data are 2D and $\mathcal{H}$ consists of linear separators. (a) The difference between two hypotheses is the probability mass under the region in which they disagree. (b) The thick line is the true hypothesis h^*, and thin lines are examples of hypotheses in $B(h^*, r)$. (c) The region of disagreement among hypotheses in $B(h^*, r)$ might look like this.

If h^* is the "true" underlying hypothesis with err$(h^*) = 0$, then after a certain amount of querying we hope to constrain the version space to be within $B(h^*, r)$ for small values of r. In that case, the probability that a random instance drawn from $\mathcal{D}$ would be queried is no more than $P_{\mathcal{D}}(\text{DIS}(B(h^*, r)))$. Under this intuition, Hanneke (2007) proposed a new parameter to characterize the complexity of active learning, as a function of the hypothesis class $\mathcal{H}$ and input distribution $\mathcal{D}$. The so-called *disagreement coefficient* ξ measures how the probability of disagreement with h^* scales with r:

$$\xi = \sup_{r>0} \frac{P_{\mathcal{D}}(\text{DIS}(B(h^*, r)))}{r}.$$

Roughly speaking, ξ attempts to quantify how quickly the region of disagreement grows as a function of the radius of the version space. Hence, smaller values of ξ imply more efficient shrinkage of the version space[1].

Theorem 6.2   Suppose $\mathcal{H}$ has finite VC dimension d, and the learning problem is separable with disagreement coefficient ξ. Then the label complexity L_{QBD} of the query by disagreement algorithm is (Hanneke, 2009):

$$L_{\text{QBD}} \leq O\left(\xi\left(d\log\xi + \log\frac{\log 1/\epsilon}{\delta}\right)\log\frac{1}{\epsilon}\right)$$

$$= \tilde{O}\left(\xi d \log\frac{1}{\epsilon}\right).$$

---

[1]A quantity called the "capacity function," which is very related to the disagreement coefficient, was independently discovered by Alexander (1987) in an analysis of passive learning problems.

Proof. Let i_t index the instance $x^{(i_t)} \frown \mathcal{D}$ sampled at the time of query t. After labeling Q queries, it holds that $|\{x^{(i_t+1)}, \ldots, x^{(i_t+Q)}\} \cap \mathrm{DIS}(\mathcal{V}_t)| \geq Q$. In other words, the set of unlabeled instances to come in from the stream that fall inside the region of disagreement is at least of size Q. Furthermore, the first Q instances in this set are drawn conditionally i.i.d. given $x \in \mathrm{DIS}(\mathcal{V}_t)$. Now, let us cleverly define Q to be:

$$Q = \left\lceil c\,\xi \left(d \log \xi + \log \frac{1}{\delta'} \right) \right\rceil.$$

Note that δ' here is not the same as δ in Theorem 6.2; we will revisit this later in the proof. For now, though, we can say that with probability $1 - \delta'$ and for all $h \in \mathcal{V}_{t+Q}$,

$$P_{\mathcal{D}}(h(x) \neq h^*(x) \mid x \in \mathrm{DIS}(\mathcal{V}_t)) \leq \frac{c'}{Q}\left(d \log \frac{Q}{d} + \log \frac{1}{\delta'} \right) \leq \frac{1}{2\xi}. \tag{6.6}$$

The first inequality comes from Theorem 6.1, by setting L equal to the passive learning bound and solving for the error tolerance ϵ. This works because the data are conditionally i.i.d. given $x \in \mathrm{DIS}(\mathcal{V}_t)$, as we saw above, and because $h(x) = h^*(x)$ for all $x \notin \mathrm{DIS}(\mathcal{V}_t)$, so all the labels outside the region of disagreement are known and available "for free." The second inequality is an algebraic simplification based on our awkward but careful definition of Q (with appropriate settings for the constants c and c').

Note that as more instances are queried and labeled, the version space necessarily shrinks. As a result, $\mathcal{V}_{t+Q} \subseteq \mathcal{V}_t$, which implies that for all $h \in \mathcal{V}_{t+Q}$:

$$
\begin{aligned}
\mathrm{err}(h) \;&=\; P_{\mathcal{D}}(h(x) \neq h^*(x)) \\[4pt]
&=\; P_{\mathcal{D}}(h(x) \neq h^*(x) \mid x \in \mathrm{DIS}(\mathcal{V}_t)) \times P_{\mathcal{D}}(x \in \mathrm{DIS}(\mathcal{V}_t)) + \\
&\quad\; P_{\mathcal{D}}(h(x) \neq h^*(x) \mid x \notin \mathrm{DIS}(\mathcal{V}_t)) \times P_{\mathcal{D}}(x \notin \mathrm{DIS}(\mathcal{V}_t)) \\[4pt]
&\leq\; \frac{P_{\mathcal{D}}(\mathrm{DIS}(\mathcal{V}_t))}{2\xi}.
\end{aligned}
$$

First, we split $\mathrm{err}(h)$ into the data proportions that fall inside and outside of the region of disagreement for $\mathcal{V}_t$. Since the data are separable, h and h^* must agree on all $x \notin \mathrm{DIS}(\mathcal{V}_t)$ by definition, so the second summand is zero. By substituting Equation 6.6 into the first summand, we arrive at an upper bound for the error, as shown in the last line. This means that $\mathcal{V}_{t+Q} \subseteq B(h^*, \frac{P_{\mathcal{D}}(\mathrm{DIS}(\mathcal{V}_t))}{2\xi})$ and that:

$$P_{\mathcal{D}}(\mathrm{DIS}(\mathcal{V}_{t+Q})) \leq P_{\mathcal{D}}\left(\mathrm{DIS}\left(B\left(h^*, \frac{P_{\mathcal{D}}(\mathrm{DIS}(\mathcal{V}_t))}{2\xi} \right) \right) \right) \leq \frac{P_{\mathcal{D}}(\mathrm{DIS}(\mathcal{V}_t))}{2}.$$

To put it another way, after $t + Q$ queries, the data density under the region of disagreement will be at most half what it was at the previous time step of t queries. So every Q queries, QBD chops the number of controversial instances in half (or better). This holds for any value of t which is a multiple of Q, and by a union bound it holds with probability $1 - \delta'\lceil\log(1/\epsilon)\rceil$ up to $t = Q\lceil\log(1/\epsilon)\rceil$. Re-substituting the definition of Q and setting $\delta' = \frac{\delta}{\log 1/\epsilon}$ results in the bound for L_{QBD} given in Theorem 6.2, which completes the proof. $\qquad\square$

Recall from Theorem 6.1 that the label complexity in the passive case was $\tilde{O}(d/\epsilon)$, while our analysis of QBD yields a label complexity of $\tilde{O}(\xi d \log \frac{1}{\epsilon})$. For well-behaved values of ξ, this implies a significant reduction in the number of labeled instances necessary to train a classifier from the same hypothesis class $\mathcal{H}$, on the same problem distribution $\mathcal{D}_{XY}$, under the same δ and ϵ requirements in the PAC setting. This is an encouraging result!

6.3 DISCUSSION

The analysis of QBD in the previous section only applies to separable (i.e., noise-free) data distributions $\mathcal{D}_{XY}$. However, the disagreement coefficient turns out to be a useful quantity for analyzing several active learning algorithms. For example, Hanneke (2007) used a proof very similar to the one above to analyze the A^2 algorithm (Balcan et al., 2006), which is designed to be able to learn from data in the "agnostic" PAC setting — that is, obtaining similar label complexity bounds in the face of arbitrarily noisy data. Dasgupta et al. (2008) also used the disagreement coefficient to analyze a different agnostic active learning algorithm, based on a reduction to supervised learning. Balcan et al. (2008) use it to show that active learning always helps under an asymptotic assumption, often with exponential improvements in sample complexity over passive learning counterparts. Beygelzimer et al. (2009) generalize the disagreement coefficient in their analysis of a novel active learning algorithm — importance-weighted active learning (IWAL) — and consider a larger family of loss functions beyond the 0/1-loss we looked at in this chapter. Interestingly, recent work has been able to place bounds on ξ for certain hypothesis classes (Friedman, 2009; Wang, 2009), which is useful for predicting exactly how much active learning algorithms might help for these classes.

For many years, the main theoretical result in active learning was an analysis of the original query by committee (QBC) algorithm by Freund et al. (1997). They showed that if the data are drawn uniformly from the surface of the unit sphere in $\mathbb{R}^K$, and the true hypothesis h^* corresponds to a homogeneous (i.e., through the origin) linear separator, then it is possible to achieve error ϵ after streaming $\tilde{O}(\frac{d}{\epsilon} \log \frac{1}{\epsilon})$ instances and querying $\tilde{O}(d \log \frac{1}{\epsilon})$ labels. Nearly a decade passed before significant progress was made toward the analysis of other active learning algorithms (or the development of novel algorithms) that could make similar claims accounting for arbitrary hypothesis classes, input distributions, and label noise. By now there are several such bounds in the literature, with a handful of the key results summarized in Table 6.1.

Still, many of these theoretical results have been based on contrived algorithms not covered in this book, which are not very practical or computationally tractable (at least not as formally defined), although there are a few exceptions (e.g., Beygelzimer et al., 2009; Dasgupta and Hsu, 2008). In fact, most analyses are limited to binary classification tasks with the goal of minimizing 0/1-loss, and are not easily adapted to other objective functions that may be more appropriate for many applications (e.g., log-loss, or squared-loss for regression tasks). Furthermore, some of the algorithms in question require an explicit enumeration over the version space, which is not only intractable for infinite hypothesis classes, but difficult to even imagine for more complex hypotheses (e.g., large heterogeneous ensembles or structured prediction models for sequences, trees, and graphs).

Table 6.1: Some key theoretical results in active learning. The last column indicates whether or not the algorithm can handle noisy labels.

Algorithm	Agnostic?
QBD: query by disagreement (Cohn et al., 1994; Hanneke, 2009)	No
QBC: query by committee (Freund et al., 1997; Seung et al., 1992)	No
A^2 algorithm (Balcan et al., 2006; Hanneke, 2007)	Yes
MBAL: margin-based AL (Balcan et al., 2007)	Yes
reduction to supervised (Dasgupta et al., 2008)	Yes
modified perceptron (Dasgupta et al., 2009)	No
IWAL: importance-weighted AL (Beygelzimer et al., 2009)	Yes

By and large, the active learning community has been split into two camps — the theoretical and the empirical — who until the last few of years have had little overlap or interaction with one another. As a result, there is still much to learn about the nature and behavior of active learning algorithms for more general and practical applications.

CHAPTER 7

Practical Considerations

"In theory, there is no difference between theory and practice. But in practice, there is."

— *Jan L.A. van de Snepscheut and/or Yogi Berra*

7.1 WHICH ALGORITHM IS BEST?

If you are working on a new problem and want to employ active learning to reduce the labeling effort involved, you clearly want to know which algorithm will work best. Unfortunately, just as knowing what feature set or learning algorithm will be most appropriate for your task (neural nets? SVMs? naive Bayes? boosted decision stumps?), there is no way to really know which active learning algorithm will be most suited to your problem *a priori*. While there have been a few large-scale empirical comparisons of some of the algorithms in this book (e.g., Körner and Wrobel, 2006; Schein and Ungar, 2007; Settles and Craven, 2008), these longitudinal studies have yielded fairly mixed results. In broad strokes, all of these algorithms seem to produce better learning curves than the passive learning baselines, although that may be an artifact of the publication bias toward positive results. There is no consistently clear-cut winner in terms of the query selection heuristics, however, which suggests that the best strategy may be dependent on the learning algorithm or application. This stresses the importance of understanding your problem first.

Table 7.1 presents a high-level comparison of the advantages and disadvantages of the various query strategy frameworks we have covered in this book. If we are mainly interested in reducing error, then error or variance reduction (Chapter 4) and hierarchical sampling (Section 5.2) are appealing since they attempt to directly optimize the training set for low generalization error. If density measures are easy to compute and there is reason to believe that cluster structure is related to the class labels, then density-weighted methods (Section 5.1) may offer a good compromise between the efficiency of simple heuristics like uncertainty or disagreement, and the value of explicit knowledge about the input distribution. If the learning algorithm is already an ensemble approach, perhaps QBC or other disagreement-based variants (Chapter 3) are most appropriate. Similarly, if using a semi-supervised learning algorithm then perhaps it would be advantageous to implement the conceptual active learning compliment (Section 5.3). For large probabilistic graphical models, which have expensive training and inference procedures, uncertainty sampling (Chapter 2) might be the only really tractable option. If you do not have the time or resources to invest in sophisticated software engineering for some of these methods, and you are a cowboy with nothing to lose, then simpler methods based on uncertainty and disagreement are probably worth at least a pilot study.

Table 7.1: A high-level comparison of the active learning strategies in this book.

Query Strategy	Advantages	Disadvantages
uncertainty sampling	simplest approach, very fast, easy to implement, usable with any probabilistic model, justifiable for max-margin classifiers	myopic, runs the risk of becoming overly confident about incorrect predictions
QBC and disagreement-based methods	reasonably simple, usable with almost any base learning algorithm, theoretical guarantees under some conditions	can be difficult to train/maintain multiple hypotheses, still myopic in terms of reducing generalization error
error/variance reduction	directly optimizes the objective of interest, empirically successful, natural extension to batch queries with some guarantees	computationally expensive, difficult to implement, limited to pool-based or synthesis scenarios, VR limited to regression models
density weighting	simple, inherits advantages of the base heuristic while making it less myopic in terms of the input distribution, can be made fast	input distribution or cluster structure may have no relationship to the labels
hierarchical sampling	exploits cluster structure, degrades gracefully if clusters are not correlated with the labels, theoretical guarantees	requires a *hierarchical* clustering of the data, which can be slow and expensive in practice, limited to pool-based scenario
active + semi-supervised	exploits latent structure in the data, aims to make good use of data through both active and semi-supervised methods	not a single algorithm/framework but a suite of approaches, inherits the pros and cons of the base algorithms

In a recent survey of annotation projects for natural language processing tasks (Tomanek and Olsson, 2009), only 20% of the respondents said they had ever decided to use active learning, which is evidence of the community's skepticism about its usefulness in practice[1]. Of the large majority who chose not to use active learning, 21% were convinced that it would not work well, with some stating that "while they believed [it] would reduce the amount of instances to be annotated, it would probably not reduce the overall annotation time." However, all but one of the eleven respondents that had attempted active learning claimed to be partially or fully satisfied with the results. In my own experience, if the nature of the problem is fairly well-understood and the appropriate learning algorithm is known (e.g., using naive Bayes, logistic regression, or support vector machines for text classification), then simple heuristics like uncertainty

[1]The authors suspect that even this is an over-estimate, since it was advertised as a survey on the use of active learning and thus biased towards those familiar with it.

sampling can work successfully. I have heard similar stories from colleagues in industry who needed to quickly build up new training sets for classifiers on a variant of a problem they were working on. If the problem is completely new, however, a safer approach is probably to conduct an initial pilot study with randomly-sampled data until the problem can be better understood, and then transition to an active learning paradigm that seems most appropriate based on preliminary results (provided there is still room for improvement!).

It should be noted that nearly all of the discussion in this book has focused on the traditional active learning problem of selecting data instances to be labeled, with certain assumptions about the cost and reliability of the "oracle" providing the labels. In many real-world deployment situations, these assumptions do not hold. Or, perhaps more interestingly, certain applications offer other opportunities such as actively soliciting rules, mixed-granularity queries, and other kinds of interactions that are much richer than the mere annotation of instances. The rest of this chapter focuses on these practical considerations, overviewing some of the work that has been done so far, and summarizing the open research questions and opportunities that are available in the rapidly-evolving landscape of active learning. Note that the following sections read more like a survey than a tutorial, and lack some details relating to these emerging research areas. The curious reader is encouraged to use the citations as a starting point for further investigation.

7.2   REAL LABELING COSTS

Throughout this book (indeed, in most active learning research) we have been motivated by reducing the number of labeled instances needed to train a good model. In essence, we assume that the *cost* (time, money, etc.) associated with labeling instances is approximately uniform, so reducing the size of $\mathcal{L}$ translates directly into cost savings. Consider structured prediction tasks, however, such as information extraction or parsing in the natural language processing literature (see Section 2.4 for an example). In these applications an annotation can be arbitrarily complex, so the time required to perform them can vary substantially. For example, longer text documents probably take a longer time to label, so the actual annotation costs differ greatly from one instance to another.

If our goal is to minimize the overall cost of training an accurate system, then simply reducing the number of labeled instances does not necessarily guarantee a reduction in cost. This motivates various approaches for *cost-sensitive active learning*. One proposed approach for reducing effort in active learning is *automatic pre-annotation*, i.e., using the current model's predictions to assist in labeling queries (Baldridge and Osborne, 2004; Culotta and McCallum, 2005). Sometimes this helps by reducing the amount of work a human must do. However, if the model makes many mistakes (which is quite possible, especially early in the process), this may also create extra work for the annotator who must correct these mistakes. To help combat this in structured prediction tasks, Culotta et al. (2006) experimented with a technique called *correction propagation*, where local edits are used to interactively update (and hopefully improve) the overall structure prediction. Nevertheless, it may also be that the active learner ends up biasing the annotator toward its poor predictions on noisy "edge" cases. For example, Baldridge and Palmer (2009) reported on a user study involving a domain novice performing rare-language glossing annotations with an active learner who did precisely this (whereas a second expert annotator was not so swayed by the model). Felt et al. (2012) found that, for a Syriac morphology-tagging task, the machine model had to achieve about 80% accuracy or more before significantly reducing the annotator's time. In general, though, automatic pre-annotation and correction propagation do not actually represent or reason about labeling costs themselves. Instead, these methods simply attempt to reduce cost indirectly by minimizing the number of labeling actions required by the human oracle.

Another group of cost-sensitive active learning approaches explicitly accounts for varied labeling costs when selecting queries. Kapoor et al. (2007) proposed a decision-theoretic approach in which the value of information includes both current labeling costs and expected future misclassification costs:

$$x^*_{VOI} = \operatorname*{argmin}_{x} \ c(x) + \mathbb{E}_{Y|\theta,x} \left[\sum_{x' \in \mathcal{U}} \kappa_{\theta+}(\hat{y}, x') \right], \tag{7.1}$$

where $c(\cdot)$ is the cost of annotating the query x, and $\kappa_{\theta+}(\cdot)$ is the cost of assigning the model's prediction $\hat{y}$ to some instance $x' \in \mathcal{U}$. The approach is very much in the same flavor as the expected error reduction framework (Chapter 4). In fact, if $c(x)$ is constant (the traditional assumption in active learning), and $\kappa_{\theta+}(\hat{y}, x') = 1 - P_{\theta+}(\hat{y}|x')$, then the equation above is equivalent to minimizing the expected future 0/1-loss (Equation 4.1). A similar derivation can be made for the expected future log-loss (Equation 4.2).

Kapoor et al. (2007) applied this approach to a voicemail classification task. Instead of using real costs, however, they made a simplifying assumption that the cost of an instance label is linear in the length of a voicemail message, and that misclassification costs are in the same currency (e.g., $c(\cdot)$ = $0.01 per second of annotation and $\kappa_{\theta+}(\cdot)$ = $10 per misclassification). This approach may not be appropriate or even straightforward for some applications, though — for example, costs may be non-deterministic, or the two cost measures may not be easily mapped into the same currency. King et al. (2004) also used this sort of approach to reduce real labeling costs. They describe a "robot scientist" which executes a series of autonomous biological experiments to discover metabolic pathways in yeast, with the objective of minimizing the cost of materials used (i.e., the cost of running an experiment, which can vary according to which lab supplies are required, plus the expected total cost of future experiments until the correct hypothesis is found). Note that instead of querying a human, this active learner is "querying nature" by conducting a laboratory experiment without a human in the loop. In this case, the cost of materials are real, but fixed and known at the time of experiment (query) selection.

An alternative approach is to select the instance that would maximize the benefit/cost ratio, or *return on investment* (ROI):

$$x^*_{ROI} \;=\; \underset{x}{\mathrm{argmax}}\; \frac{\phi_A(x) - c(x)}{c(x)}$$

$$\;=\; \underset{x}{\mathrm{argmax}}\; \frac{\phi_A(x)}{c(x)} \;, \tag{7.2}$$

where $\phi_A(x)$ is the benefit, i.e., any base utility measure (entropy-based uncertainty sampling, for example). The simplification on the second line comes from splitting the fraction and ignoring the resulting cost constant. A significant advantage of this approach is that we do not necessarily have to map utilities and costs into the same currency, as we would have to for labeling and (mis)classification costs in Equation 7.1.

In all the settings above, and indeed in most of the cost-sensitive active learning literature, the cost of annotating an instance is assumed to be fixed and known to the learner before querying. In some settings, annotation costs are variable and not known *a priori*, for example, when labeling cost is a function of elapsed annotation time. Settles et al. (2008a) addressed this by training a regression cost-model (alongside the active task-model) which tries to learn how to predict the real, unknown annotation cost using a few simple "meta features" on instances. An analysis of several user studies using real human annotation costs revealed the following:

- In some domains, annotation costs are not (approximately) constant across instances, and can indeed vary considerably. This result is supported by the subsequent findings of others, working on different learning tasks (Arora et al., 2009; Haertel et al., 2008; Vijayanarasimhan and Grauman, 2009a; Wallace et al., 2010b).

- Consequently, active learning approaches which ignore cost may perform no better than random selection (i.e., passive learning).

- The cost of annotating an instance may not be intrinsic, but may instead vary based on the *person* doing the annotation. This result is also supported by the findings of Ringger et al. (2008) and Arora et al. (2009).

- The measured cost for an annotation may include stochastic components. In particular, there are at least two types of noise which affect annotation speed: *jitter* (minor variations due to annotator fatigue, latency, etc.) and *pause* (unexpected major variations, such as a phone call or other interruption).

- Unknown annotation costs can *sometimes* be accurately predicted, even after seeing only a few labeled training instances. This result is also supported by the findings of others (Vijayanarasimhan and Grauman, 2009a; Wallace et al., 2010b). Moreover, these learned cost-models are more accurate than simple cost heuristics (e.g., a linear function of document length).

Several empirical studies have shown that learned cost models can be trained to predict reasonably accurate annotation times   (Haertel et al., 2008; Settles et al., 2008a; Vijayanarasimhan and Grauman, 2009a; Wallace et al., 2010b). However, there are mixed results as to how these predicted costs can be effectively incorporated into a query selection algorithm. Settles et al. (2008a) found that using ROI (7.2) with predicted costs is not necessarily any better than random sampling for several natural language tasks (this is true even if *actual* costs are known). However, results from Donmez and Carbonell (2008), Haertel et al. (2008), and Wallace et al. (2010b) suggest that ROI is sometimes effective, even when using a learned cost model. Tomanek and Hahn (2010) evaluated the effectiveness or ROI using the least confident uncertainty sampling strategy $\phi_{LC}(\cdot)$ (Equation 2.1), and found that it worked better with the non-linear transformation $\exp(\beta\phi_{LC}(\cdot))$, using some scaling parameter β. This suggests that ROI requires an appropriate relationship between the benefit and cost functions to work well. Vijayanarasimhan and Grauman (2009a) demonstrated potential cost savings in active learning using predicted annotation costs in a computer vision task using a decision-theoretic approach based on Equation 7.1. It is unclear whether these mixed results are intrinsic, task-specific, or simply a result of differing experimental assumptions. Therefore, there are still many open questions as to how we can effectively use real and predicted annotations costs in practical settings.

7.3   ALTERNATIVE QUERY TYPES

Another way to extend the traditional notion of active learning and make them more suitable for use in practice is to reconsider the kinds of queries that can be asked. A traditional active learner only queries instances to be labeled by the oracle. What other forms might a query take?

There has been considerable interest in recent years in how to incorporate human *domain knowledge* into machine learning algorithms. Think of it this way: the learner aims to induce hypotheses both from human advice (e.g., rules, constraints, or other forms of domain knowledge) and data (e.g., both labeled and unlabeled instances). Such hybrid approaches of "explanation-based" and empirical, statistical machine learning have a long history, beginning in the mid-1990s with neural networks (Towell and Shavlik, 1994), and continuing through modern learning systems like SVMs (Kunapuli et al., 2011; Small et al., 2011), topic models (Andrzejewski et al., 2011), and even more complex graphical models (Ganchev et al., 2010; Mann and McCallum, 2010). It is natural, then, to think about active learning algorithms that can formulate and pose queries in the language of rules and advice, in addition to querying unlabeled instances.

A very simple form of active learning for domain knowledge is to solicit information about features. Raghavan et al. (2006) proposed one such approach, *tandem learning*, which incorporates feature feedback in traditional classification problems. In their work, a text classifier may interleave instance-label queries with feature-salience queries (e.g., "is the word *puck* a discriminative feature for classifying sports documents?"). Values for the salient features are then amplified in instance feature vectors to reflect their relative importance. Raghavan et al. reported that interleaving such queries is very effective for text classification, and also found that words (or features) are often much easier for human annotators to label in empirical user studies. Note, however, that these "feature labels" imply their discriminative value and do not tie features to class labels directly.

An extension of this idea is active learning by labeling features, sometimes called *active dual supervision*. In this setting, the oracle may label features which are deemed to be good predictors of one or more classes. There are several ways of incorporating these feature labels into the model, e.g., using priors on model parameters (Settles, 2011), mixture models induced from rules as well as data (Attenberg et al., 2010), label propagation in graph-based learning algorithms (Sindhwani et al., 2009), or constraints on the objective function (Druck et al., 2008; Liang et al., 2009; Small et al., 2011). As an example of constraints on the objective function, consider a scenario where users can specify rules relating features to label distributions, e.g., "95% of the time, when the word *puck* is observed in a document, the class label is hockey." The learning algorithm then tries to find a set of model parameters that match expected label distributions over the unlabeled pool $\mathcal{U}$ against these user-specified priors. In practice, people do a poor job of estimating these kinds of distributions (Tversky and Kahneman, 1983). However, the learning algorithm can be made robust to these errors by treating them as soft constraints that do not need to be matched exactly, if the data disagree or multiple rules are in conflict. In fact, Mann and McCallum (2008) found that specifying many imprecise constraints tends to result in better models than fewer more precise constraints, suggesting that human-specified feature labels (however noisy) are useful if there are enough of them. This begs the question of how to *actively* solicit these kinds of feature labels.

Druck et al. (2009) proposed and evaluated several active query strategies aimed at gathering useful feature labels. They show that active feature labeling is more effective than either "passive" feature labeling (using a variety of strong baselines) or instance-labeling (both passive and active) for two information extraction tasks. These results held true for both simulated experiments and real, interactive human user studies. The query algorithm that proved most effective was a form of density-weighted uncertainty sampling (Section 5.1), but for features:

$$ f_{WU}^* \;=\; \underset{f}{\mathrm{argmax}}\; \mathbb{E}_{\mathcal{U}}\left[H_\theta(Y|f)\right] \times \log C_f, $$

where f is a candidate feature, $\mathbb{E}_{\mathcal{U}}[\cdot]$ is the expectation over the unlabeled input distribution, and C_f is the frequency of the feature f in the corpus. The first term is the average uncertainty of items which contain that feature (in this work, features were discrete binary variables like words, capitalization patterns, etc.) times a scaling factor to represent how often the feature occurs, and thus how much impact its label is likely to have. One explanation for the excellent gains in user studies is that features are often both intuitively easier to label (requiring 2–4 seconds in user studies, compared to arbitrarily long amounts of time for instances, depending on the task) and provide much more information: a feature label is a rule which covers all instances containing that feature, whereas an instance label only provides information about the instance in question.

Settles (2011) built on this work in a slightly different way for an interactive text classification system called DUALIST. Figure 7.1 shows the interactive annotation interface applied to a sentiment analysis task: identifying positive and negative movie reviews (Pang et al., 2002). Instances are text

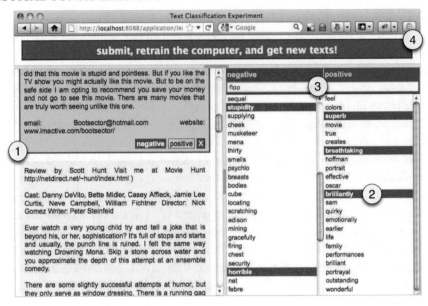

Figure 7.1: User interface for the DUALIST system (Settles, 2011).

documents, and features are words. At any time, the user may choose among four annotation actions: ① label document queries, ② label word queries, ③ volunteer labeled words, or ④ submit all labels, retrain the classifier, and obtain the next set of queries. The differences compared to previous work are threefold. First, instance and feature queries are interleaved. Second, the human oracle is allowed to volunteer his or her knowledge about feature labels by typing them into text boxes, so not all annotations are at the behest of the learner's requests. Third, the feature queries are organized into columns according to label, which not only aims to reduce cognitive load for the annotator, but requires some "initiative" by the active learner to both identify informative features, and to try and postulate the associated label(s)[2]. The query selection strategy in this case is the expected information gain of a feature:

$$f_{EIG}^* = \operatorname*{argmax}_f \ \mathbb{E}_{\mathcal{U}}\left[H_\theta(Y)\right] - \mathbb{E}_{\mathcal{U}}\left[H_\theta(Y|f)\right],$$

that is, the difference between the model's average uncertainty of all the unlabeled instances, and the uncertainty of the instances with the feature present. This is the expected amount of information that feature f provides about the label variable Y. In a similar vein, features can be assigned to class label columns according to the expected frequency with which f appears in instances of class y according to the model's predictions. In user studies on several benchmark data sets, this interactive approach outperformed both active and passive learning by labeling instances (documents) alone.

---

[2]For tasks with more than two classes, features may take on multiple labels. For example, the word "goalie" might imply both hockey and soccer documents, but not baseball or tennis, etc.

bag: image = { instances: segments }

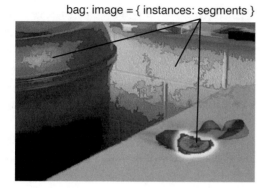

bag: document = { instances: passages }

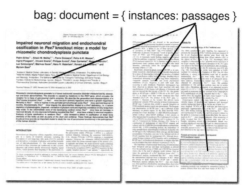

(a) content-based image classification

(b) text classification

Figure 7.2: Multiple-instance active learning. (a) Images can be represented as "bags" and instances correspond to segments. An MI active learner may directly query which segments belong to an object of interest, such as the gold medal here. (b) Documents are represented as bags and instances are passages of text. An MI learner may query specific passages to determine if they represent the class of interest, alleviating the need to read the entire document.

We can also think of interacting with the learner at a granularity somewhere in between features (very fine) and instances (very coarse). Settles et al. (2008b) introduced a scenario called *multiple-instance active learning*, illustrated in Figure 7.2 for (a) computer vision and (b) natural language processing tasks. In multiple-instance (MI) learning, instances are grouped into *bags* (i.e., multi-sets), and it is the bags, rather than instances, that are labeled for training. A bag is labeled negative if and only if all of its instances are negative. A bag is labeled positive, however, if at least one of its instances is positive (note that positive bags may also contain negative instances). A naive approach to MI learning is to view it as supervised learning with one-sided noise (i.e., all negative instances are truly negative, but some positives are actually negative). However, special MI learning algorithms have been developed to learn from labeled bags despite this ambiguity. The MI setting was formalized by Dietterich et al. (1997) in the context of drug activity prediction, and has since been applied to a wide variety of tasks including content-based image retrieval (Andrews et al., 2003; Maron and Lozano-Perez, 1998; Rahmani and Goldman, 2006) and text classification (Andrews et al., 2003; Ray and Craven, 2005).

An advantage of the MI representation in Figure 7.2(a) is that it may be much easier to obtain image labels than it is to get fine-grained segment labels. Similarly, for texts classification in Figure 7.2(b), document labels may be more readily available than passages (e.g., paragraphs). The MI representation is compelling for classification tasks for which bag labels are freely available or cheaply obtained (e.g., from online indexes and databases), but the target concept is represented by only a few portions (i.e., image segments or passages of text). It is often possible to obtain labels at both levels of granularity. Fully labeling all instances, however, would be expensive: the rationale for an MI representation is often that it allows us to take advantage of coarse labelings that may be available at low cost (or even for free). In MI active learning, however, the learner is sometimes allowed to query for labels at a finer granularity than the target concept, e.g., querying passages rather than entire documents, or segmented image regions rather than entire images. Settles et al. (2008b) focused on this type of mixed-granularity active learning with a multiple-instance generalization of logistic regression and develop weighted uncertainty sampling algorithms that also account for how important an instance is to the MI classifier. Vijayanarasimhan and Grauman (2009a,b) extended the idea to SVMs for the image retrieval task, and also explore an approach that interleaves queries at varying levels of granularity and cost, using a decision-theoretic approach similar to Equation 7.1.

7.4   SKEWED LABEL DISTRIBUTIONS

In many practical applications of machine learning, the distribution of labels is not very balanced. For example, if we want to train a classifier to identify "interesting" web pages in some sense, then the pages of interest in our crawl might be outnumbered by a million to one by pages that are not of interest. When labels are skewed in this way, active learning may offer little or no gain over random sampling, and the problem is not merely due to learning with skewed distributions. The lack of labeled data exacerbates the problem because the active learning strategies may not be able to focus on the interesting minority instances of which the learner is unaware.

Figure 7.3 illustrates this problem for the task of classifying web pages with various amounts of artificially-induced label skew. As the positive class becomes more and more rare, random sampling rapidly degrades. Uncertainty sampling remains somewhat more robust up to a point, but with a moderately severe skew (10,000 to one), it offers no gains. To combat this problem, Attenberg and Provost (2010a) proposed "guided learning," which allows a human annotator to explicitly search for class-representative instances as opposed to (or in addition to) labeling specific queries or random samples. They found that guided learning can outperformed active learning in terms of real annotation costs as long as searching for these class-representative instances was less than eight times more expensive (per instance) than labeling query instances. Figure 7.3 shows that this approach can help to maintain good quality ranking and classifications in the face of extreme data imbalance.

Arguably, though, this is simply due to a bad match between the selection strategy and the desired objective function. Recall from Chapter 6 that uncertainty sampling is a crude approximation to a utility measure that maximizes expected information gain (i.e., reducing prediction uncertainty). The evaluation in Figure 7.3 is measured instead against area under the ROC curve (AUC), which is the probability of ranking a randomly-chosen **positive** instance above a randomly-chosen **negative**. Although computationally costly, a decision-theoretic utility function that directly maximizes the expected AUC of the minority class may perform better that uncertainty sampling here.

Wallace et al. (2010a) successfully used a dual supervision approach (Section 7.3) with active learning under extreme class imbalance for biomedical citation screening. In this case, they built

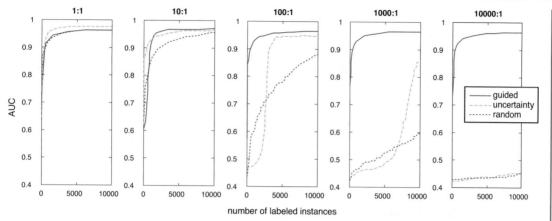

Figure 7.3: The effects of artificially-induced label skew on a text classification task (sports vs. non-sports web pages). Learning curves show area under the ROC curve (AUC) of logistic regression as a function of the number of labeled documents. As the positive class becomes more and more rare, uncertainty learning remains more robust than random sampling, but with a severe skew (10,000 to one) there are virtually no gains. "Guided" learning, by which the oracle is allowed to search and provide labels for representative data instances, helps in this respect. *Source:* Adapted from Attenberg and Provost (2010b), used with kind permission of the authors.

two models: one from labeled data in the normal supervised way, and another based on labeled words, with the assumption that these feature labels cover the space of representative clusters. They used query by disagreement (Section 3.3) between these two models to perform active learning on additional documents for querying, and found that this dual-view strategy helps to correct some of the problems of label skew compared to several standard active learning baselines.

7.5   UNRELIABLE ORACLES

Another strong assumption in most active learning work is that the quality of labeled data is high. However, if labels come from an empirical experiment (e.g., in biological, chemical, or clinical studies), then one can usually expect some noise to result from the instrumentation or experimental setting. Even if labels come from human experts, they may not always be reliable, for several reasons. First, some instances are implicitly difficult for both people and machines (see Section 7.2), and second, people can become distracted or fatigued over time, introducing variability in the quality of their annotations. The recent introduction of Internet-based "crowdsourcing" tools such as Amazon's Mechanical Turk[3] and the clever use of online annotation games[4] have enabled some researchers to attempt to "average out" some of this noise by cheaply obtaining labels from multiple non-

[3]http://www.mturk.com
[4]http://www.gwap.com

experts. Such approaches have been used to produce gold-standard quality training sets (Snow et al., 2008) and also to evaluate learning algorithms on data for which no gold-standard labelings exist (Carlson et al., 2010; Mintz et al., 2009).

The question remains about how to use non-experts (or even noisy experts) as oracles in active learning. In particular, when should the learner decide to query for the (potentially noisy) label of a *new* unlabeled instance, versus querying for repeated labels to de-noisify an *existing* training instance that seems a bit suspect? Sheng et al. (2008) studied this problem using several heuristics that take into account estimates of both oracle and model uncertainty, and show that data can be improved by selective repeated labeling. Their analysis assumes that: (i) all oracles are equally and consistently noisy; and (ii) annotation is a noisy process over some underlying true label. Donmez et al. (2009) addressed the first issue by allowing annotators to have different noise levels, and show that both true instance labels and individual oracle qualities can be estimated (so long as they do not change over time). They take advantage of these estimates by querying only the more reliable annotators in subsequent iterations active learning. In follow-up work, Donmez et al. (2010) used a particle filter to deal with noisy oracles whose quality varies over time (e.g., improving after becoming more familiar with the task, or degrading after becoming fatigued). Wallace et al. (2011) also investigated methods for active learning with annotators of varying levels of expertise. In particular, they assume that novice or non-expert annotators (e.g., crowdsourced users) are able to estimate the quality of their own labels and indicate when instances were too difficult. A small user study confirmed that this approach made better use of both novices and experts.

There are still many open research questions along these lines. For example, how might the effect of payment influence annotation quality (i.e., if you pay a non-expert twice as much, are they likely to try and be more accurate)? What if some instances are inherently noisy regardless of which oracle is used, and repeated labeling is not likely to improve matters? Finally, in most crowdsourcing environments the users are not necessarily available "on demand," thus accurate estimates of annotator quality may be difficult to achieve in the first place, and might possibly never be applicable again, since the model has no real choice over which oracles to use. How might the learner continue to make progress?

7.6   MULTI-TASK ACTIVE LEARNING

The typical active learning setting assumes a single machine learner trying to solve a single task. In many real-world problems, however, the same data might be labeled in multiple ways for several different subtasks. In such cases, it is probably more economical to label a single instance for all subtasks simultaneously, or to choose instance-task query pairs that provide as much information as possible to *all* tasks. This motivates the need for *multi-task active learning* algorithms. If we take a multi-task entropy-based uncertainty sampling sort of approach, then we might want to select instances with the highest joint conditional entropy of both labels given the instance: $H_\theta(Y_1, Y_2|x)$, where Y_1 and Y_2 denote the output variables for the two different tasks.

The relationship among different subtasks can be a major factor in how to design a multi-task active learner. For example, imagine a multi-task text classification problem where the two tasks are arranged in a pipeline: Y_1 denotes the language of a text (english, french, portuguese, etc.) and Y_2 denotes the sentiment (positive or negative). The second task probably depends on the first task, i.e., we want to train a separate sentiment model for each language. In this case, our query strategy decomposes quite nicely:

$$
\begin{aligned}
x^*_{MT} &= \underset{x}{\operatorname{argmax}} \ H_\theta(Y_1, Y_2|x) \\[2mm]
&= \underset{x}{\operatorname{argmax}} \ H_\theta(Y_1|x) + H_\theta(Y_2|Y_1, x) \\[2mm]
&= \underset{x}{\operatorname{argmax}} \ H_\theta(Y_1|x) + \mathbb{E}_{Y_1|x}\big[H_\theta(Y_2|y_1, x)\big].
\end{aligned}
$$

The joint utility measure for our two tasks is simply the sum of the language model uncertainty, plus the sentiment model uncertainty as an expectation over the languages. As we saw in Equation 6.3, this is an approximation to reducing the expected future joint log-loss for the two tasks.

In some cases, the labels may be inherently linked to one another by *output constraints* which might be known ahead of time. For example, imagine a system that classifies encyclopedia articles into hundreds of categories that happen to be arranged in a hierarchy. We may train a separate classifier for each category (node in the hierarchy), but because of the hierarchy we know that an article about a biologist is also an article about a scientist and therefore a person. So a positive label for biologist implies that both scientist and person are positive as well, and also implies that sportsTeam and fruit are false (since they lie in parts of the ontology that are mutually exclusive with biologist). Entropy-based uncertainty sampling for a single task selects instances with the highest expected log-loss:

$$
\begin{aligned}
x^*_H &= \underset{x}{\operatorname{argmax}} \ H_\theta(Y|x) \\[2mm]
&= \underset{x}{\operatorname{argmax}} \ \mathbb{E}_{Y|x}\big[-\log P_\theta(y|x)\big].
\end{aligned}
$$

However, if there are many tasks with output constraints, we can generalize this to the expected total log-loss of all labelings that are implied by the label of a task in question. Imagine that the active learner can choose an instance-task query pair $\langle x, t \rangle$. This more generalized utility measure can be written as:

$$
\langle x, t \rangle^*_{MT} = \underset{x,t}{\operatorname{argmax}} \ \mathbb{E}_{Y_t|x}\left[-\sum_{y_{t'} \Leftarrow y_t} \log P_\theta(y_{t'}|x)\right],
$$

where the summand inside the expectation iterates over all task labelings $y_{t'}$ that are implied by the labeling y_t (e.g., {biologist, scientist, person, ¬sportsTeam, ¬fruit} ⟸ biologist). Zhang (2010) used this sort of approach for a few simple cross-task information extraction and text classification experiments with successful results. Qi et al. (2008) considered a slightly different version of multi-task active learning with output dependencies in a computer vision problem, where the output constraints are *not* necessarily know a priori.

Finally, it may be that the two tasks are not necessarily related at all, or the interdependencies are too complex to model explicitly. Reichart et al. (2008) experimented with two methods for such settings. The first is *alternating selection*, which allows learner predicting Y_1 to query instances in one iteration of active learning — using an appropriate query selection scheme, uncertainty sampling in their case — and let the Y_2 learner query instances in the next iteration, and so on. The approach can be generalized to more than two tasks using a round-robin approach (possibly stochastic). The second method is *rank combination*, which lets all the learners independently rank-order queries from the pool, and instances with the lowest combined rank are selected for querying. They experiment with these approaches to select sentences for two natural language tasks (parsing and information extraction), and conclude that they produce better learning curves than random sampling. However, each subtask's curves were not as high as active learning for that subtask alone, which (not surprisingly) came at the expense of the other task. Thus, the combined approach was preferable on average across all the tasks. Roth and Small (2008) also proposed an adaptive strategy for pipeline models which is similar to rank combination. Specifically, their approach selects instances based on a weighted combination of utility measures from each stage in the pipeline.

7.7   DATA REUSE AND THE UNKNOWN MODEL CLASS

The labeled training set $\mathcal{L}$ collected by an active learner nearly always results in a biased distribution. This bias is implicitly tied to the class of model being used to select the queries. For example, the labeled instances in Figure 2.2(c) are clustered around a linear decision boundary. Frequently in machine learning, as the state of the art advances we want to switch to a new and improved learner. This can be a problem if we wish to "reuse" this training data with models of a different type, or if we do not even know the appropriate model class (or feature set) for the task to begin with. Nothing is really known theoretically about the ability to reuse labeled data gathered by one learner for another, and the empirical body of work is limited. Tomanek and Morik (2011) presented a large empirical study of sample reuse among many different models for benchmark classification and information extraction tasks, but the results (like previous work) are somewhat inconclusive.

Lewis and Catlett (1994) found that decision tree classifiers can still benefit from a training set constructed by an active naive Bayes learner using uncertainty sampling. Tomanek et al. (2007) also showed that information extraction data gathered by a logistic regression classifier using QBC can be effectively reused to train linear-chain conditional random fields for information extraction tasks, maintaining cost savings compared with random sampling. Hwa (2001) successfully reused natural language parsing data selected by one type of parser to train other types of parsers. However, Baldridge and Osborne (2004) encountered the exact opposite problem when re-using data selected by one parsing model to train a variety of other parsers. As an alternative, they perform active learning using a heterogeneous ensemble composed of different parser types, and also use semi-automated labeling to cut down on human annotation effort (see Section 7.2). This approach helped to reduce the amount of work required for each parser type compared with passive learning. Similarly, Lu and Bongard (2009) employed active learning with a heterogeneous ensemble of neural networks and decision trees, when the more appropriate model was not known in advance. Their ensemble approach is able to simultaneously select informative instances for the overall model, as well as bias the constituent weak learners toward the more appropriate model class as it learns. Sugiyama and Rubens (2008) experimented with an ensemble of linear regression models using different feature sets, to study cases in which the appropriate feature set is not yet decided upon.

This highlights a very important issue for active learning in practice. If the best model class and feature set happen to be known in advance — or if these are not likely to change much in the future — then active learning can possibly be safely used. Otherwise, random sampling (at least for pilot studies, until the task can be better understood) may be more advisable than taking one's chances on active learning with an inappropriate model. One viable active approach seems to be the use of heterogeneous ensembles in selecting queries, but there is still much work to be done in this direction.

7.8   STOPPING CRITERIA

A potentially important element of interactive learning applications in general is knowing when to *stop* learning, or at least to stop posing queries. The best way to think about this is the point at which the cost of acquiring new training data is greater than the cost of the errors made by the current system. In other words, we want to recognize when the accuracy of a learner has reached a plateau, and acquiring more data is likely a waste of resources. Since active learning is concerned with improving accuracy while remaining sensitive to data acquisition costs, it is natural to think about devising a "stopping criterion" for active learning, i.e., a method by which an active learner may decide to stop asking questions in order to conserve resources.

Several such stopping criteria for active learning have been proposed (Bloodgood and Shanker, 2009; Olsson and Tomanek, 2009; Vlachos, 2008). These methods are all fairly similar, generally based on the notion that there is an intrinsic measure of stability or self-confidence within the learner, and active learning ceases to be useful once that measure begins to level-off or degrade. Such self-stopping methods seem like a good idea, and may be applicable in certain situations. However, in my own experience, the real stopping criterion for practical applications is based on economic or other external factors, which likely come well before an intrinsic learner-decided threshold.

APPENDIX A

Nomenclature Reference

"What's the use of their having names," the Gnat said, "if they won't answer to them?"

"No use to them," said Alice; "but it's useful to the people who name them, I suppose. If not, why do things have names at all?"

— *Lewis Carroll*

$\mathcal{L}$	labeled data set
$\mathcal{U}$	unlabeled data set
x, y	input data instance and corresponding label
$\mathcal{H}$	hypothesis space (i.e., model class)
$\mathcal{V}$	version space (i.e., subset of $\mathcal{H}$ consistent with $\mathcal{L}$)
h	hypothesis
θ	parameter(s) of a particular hypothesis or model
$\phi_A(\cdot)$	utility measure A
x_A^*	best query instance according to $\phi_A(\cdot)$
$\text{DIS}(\cdot)$	region of disagreement
ξ	disagreement coefficient
$H(\cdot)$	entropy
$KL(\cdot \| \cdot)$	Kullback-Liebler (KL) divergence
$I(\cdot; \cdot)$	mutual information / information gain
∇x	Fisher score
F	Fisher information matrix

Bibliography

N. Abe and H. Mamitsuka. Query learning strategies using boosting and bagging. In *Proceedings of the International Conference on Machine Learning (ICML)*, pages 1–9. Morgan Kaufmann, 1998. Cited on page(s) 29

K.S. Alexander. Rates of growth and sample moduli for weighted empirical processes indexed by sets. *Probability Theory and Related Fields*, 75(3):379–423, 1987. DOI: 10.1007/BF00318708 Cited on page(s) 59

S. Andrews, I. Tsochantaridis, and T. Hofmann. Support vector machines for multiple-instance learning. In *Advances in Neural Information Processing Systems (NIPS)*, volume 15, pages 561–568. MIT Press, 2003. Cited on page(s) 71

D. Andrzejewski, X. Zhu, M. Craven, and B. Recht. A framework for incorporating general domain knowledge into latent dirichlet allocation using first-order logic. In *Proceedings of the International Joint Conference on Artificial Intelligence (IJCAI)*, pages 1171–1177. AAAI Press, 2011. DOI: 10.5591/978-1-57735-516-8/IJCAI11-200 Cited on page(s) 68

D. Angluin. Queries and concept learning. *Machine Learning*, 2:319–342, 1988. DOI: 10.1023/A:1022821128753 Cited on page(s) 6, 32

D. Angluin. Queries revisited. In *Proceedings of the International Conference on Algorithmic Learning Theory*, pages 12–31. Springer-Verlag, 2001. DOI: 10.1007/3-540-45583-3_3 Cited on page(s) 6

S. Arora, E. Nyberg, and C.P. Rosé. Estimating annotation cost for active learning in a multi-annotator environment. In *Proceedings of the NAACL HLT Workshop on Active Learning for Natural Language Processing*, pages 18–26. ACL, 2009. DOI: 10.3115/1564131.1564136 Cited on page(s) 67

L. Atlas, D. Cohn, R. Ladner, M. El-Sharkawi, R. Marks II, M. Aggoune, and D. Park. Training connectionist networks with queries and selective sampling. In *Advances in Neural Information Processing Systems (NIPS)*, volume 3, pages 566–573. Morgan Kaufmann, 1990. Cited on page(s) 7

J. Attenberg and F. Provost. Why label when you can search? alternatives to active learning for applying human resources to build classification models under extreme class imbalance. In *Proceedings of the International Conference on Knowledge Discovery and Data Mining (KDD)*, pages 423–432. ACM, 2010a. DOI: 10.1145/1835804.1835859 Cited on page(s) 72

J. Attenberg and F. Provost. Inactive learning? difficulties employing active learning in practice. *SIGKDD Explorations*, 12(2):36–41, 2010b. DOI: 10.1145/1964897.1964906 Cited on page(s) 73

J. Attenberg, P. Melville, and F. Provost. A unified approach to active dual supervision for labeling features and examples. In *Proceedings of the European Conference on Machine Learning and Principles and Practice of Knowledge Discovery in Databases (ECML PKDD)*, pages 40–55. Springer, 2010. DOI: 10.1007/978-3-642-15880-3_9 Cited on page(s) 69

M.F. Balcan, A. Beygelzimer, and J. Langford. Agnostic active learning. In *Proceedings of the International Conference on Machine Learning (ICML)*, pages 65–72. ACM, 2006. DOI: 10.1145/1143844.1143853 Cited on page(s) 61, 62

M.F. Balcan, A. Broder, and T. Zhang. Margin based active learning. In *Proceedings of the Conference on Learning Theory (COLT)*, pages 35–50. Springer, 2007. Cited on page(s) 62

M.F. Balcan, S. Hanneke, and J. Wortman. The true sample complexity of active learning. In *Proceedings of the Conference on Learning Theory (COLT)*, pages 45–56. Springer, 2008. Cited on page(s) 61

J. Baldridge and M. Osborne. Active learning and the total cost of annotation. In *Proceedings of the Conference on Empirical Methods in Natural Language Processing (EMNLP)*, pages 9–16. ACL, 2004. Cited on page(s) 65, 76

J. Baldridge and A. Palmer. How well does active learning *actually* work? Time-based evaluation of cost-reduction strategies for language documentation. In *Proceedings of the Conference on Empirical Methods in Natural Language Processing (EMNLP)*, pages 296–305. ACL, 2009. DOI: 10.3115/1699510.1699549 Cited on page(s) 65

A.L. Berger, V.J. Della Pietra, and S.A. Della Pietra. A maximum entropy approach to natural language processing. *Computational Linguistics*, 22(1):39–71, 1996. Cited on page(s) 39

A. Beygelzimer, S. Dasgupta, and J. Langford. Importance-weighted active learning. In *Proceedings of the International Conference on Machine Learning (ICML)*, pages 49–56. Omnipress, 2009. DOI: 10.1145/1553374.1553381 Cited on page(s) 61, 62

C.M. Bishop. *Pattern Recognition and Machine Learning*. Springer, 2006. DOI: 10.1117/1.2819119 Cited on page(s) xi

D.M. Blei, A.Y. Ng, and M. Jordan. Latent dirichlet allocation. *Journal of Machine Learning Research*, 3:993–1022, 2003. Cited on page(s) 52

M. Bloodgood and V. Shanker. A method for stopping active learning based on stabilizing predictions and the need for user-adjustable stopping. In *Proceedings of the Conference on Natural Language*

Learning (CoNLL), pages 39–47. ACL, 2009. DOI: 10.3115/1596374.1596384 Cited on page(s) 77

A. Blum and T. Mitchell. Combining labeled and unlabeled data with co-training. In *Proceedings of the Conference on Learning Theory (COLT)*, pages 92–100. Morgan Kaufmann, 1998. DOI: 10.1145/279943.279962 Cited on page(s) 53

L. Breiman. Bagging predictors. *Machine Learning*, 24(2):123–140, 1996. DOI: 10.1023/A:1018094028462 Cited on page(s) 28, 30, 38

K. Brinker. Incorporating diversity in active learning with support vector machines. In *Proceedings of the International Conference on Machine Learning (ICML)*, pages 59–66. AAAI Press, 2003. Cited on page(s) 44

R. Burbidge, J.J. Rowland, and R.D. King. Active learning for regression based on query by committee. In *Proceedings of Intelligent Data Engineering and Automated Learning (IDEAL)*, pages 209–218. Springer, 2007. DOI: 10.1007/978-3-540-77226-2_22 Cited on page(s) 31

A. Carlson, J. Betteridge, R. Wang, E.R. Hruschka Jr, and T. Mitchell. Coupled semi-supervised learning for information extraction. In *Proceedings of the International Conference on Web Search and Data Mining (WSDM)*, pages 101–110. ACM, 2010. DOI: 10.1145/1718487.1718501 Cited on page(s) 74

K. Chaloner and I. Verdinelli. Bayesian experimental design: A review. *Statistical Science*, 10(3): 237–304, 1995. DOI: 10.1214/ss/1177009939 Cited on page(s) 41, 42

D. Cohn. Neural network exploration using optimal experiment design. In *Advances in Neural Information Processing Systems (NIPS)*, volume 6, pages 679–686. Morgan Kaufmann, 1994. DOI: 10.1016/0893-6080(95)00137-9 Cited on page(s) 43

D. Cohn, L. Atlas, and R. Ladner. Improving generalization with active learning. *Machine Learning*, 15(2):201–221, 1994. DOI: 10.1007/BF00993277 Cited on page(s) 7, 8, 24, 58, 62

D. Cohn, Z. Ghahramani, and M.I. Jordan. Active learning with statistical models. *Journal of Artificial Intelligence Research*, 4:129–145, 1996. Cited on page(s) 6, 43

T. H. Corman, C. E. Leiserson, and R. L. Rivest. *Introduction to Algorithms*. MIT Press, 1992. Cited on page(s) 17

T.M. Cover and J.A. Thomas. *Elements of Information Theory*. Wiley, 2006. Cited on page(s) 41

A. Culotta, T. Kristjansson, A. McCallum, and P. Viola. Corrective feedback and persistent learning for information extraction. *Artificial Intelligence*, 170:1101–1122, 2006. DOI: 10.1016/j.artint.2006.08.001 Cited on page(s) 65

A. Culotta and A. McCallum. Reducing labeling effort for stuctured prediction tasks. In *Proceedings of the National Conference on Artificial Intelligence (AAAI)*, pages 746–751. AAAI Press, 2005. Cited on page(s) 17, 65

I. Dagan and S. Engelson. Committee-based sampling for training probabilistic classifiers. In *Proceedings of the International Conference on Machine Learning (ICML)*, pages 150–157. Morgan Kaufmann, 1995. Cited on page(s) 8, 28

A. Das and D. Kempe. Algorithms for subset selection in linear regression. In *Symposium on Theory of Computing (STOC)*, pages 45–54. ACM, 2008. DOI: 10.1145/1374376.1374384 Cited on page(s) 45

S. Dasgupta. Two faces of active learning. *Theoretical Computer Science*, 412(19):1767–1781, 2010. DOI: 10.1016/j.tcs.2010.12.054 Cited on page(s) xiii

S. Dasgupta and D.J. Hsu. Hierarchical sampling for active learning. In *Proceedings of the International Conference on Machine Learning (ICML)*, pages 208–215. ACM, 2008. DOI: 10.1145/1390156.1390183 Cited on page(s) 50, 52, 61

S. Dasgupta, D. Hsu, and C. Monteleoni. A general agnostic active learning algorithm. In *Advances in Neural Information Processing Systems (NIPS)*, volume 20, pages 353–360. MIT Press, 2008. Cited on page(s) 61, 62

S. Dasgupta, A.T. Kalai, and C. Monteleoni. Analysis of perceptron-based active learning. *Journal of Machine Learning Research*, 10:281–299, 2009. DOI: 10.1007/11503415_17 Cited on page(s) 62

V.R. de Sa. Learning classification with unlabeled data. In *Advances in Neural Information Processing Systems (NIPS)*, volume 6, pages 112–119. MIT Press, 1994. Cited on page(s) 53

T. Dietterich, R. Lathrop, and T. Lozano-Perez. Solving the multiple-instance problem with axis-parallel rectangles. *Artificial Intelligence*, 89:31–71, 1997. DOI: 10.1016/S0004-3702(96)00034-3 Cited on page(s) 71

P. Donmez and J. Carbonell. Proactive learning: Cost-sensitive active learning with multiple imperfect oracles. In *Proceedings of the Conference on Information and Knowledge Management (CIKM)*, pages 613–622. ACM, 2008. DOI: 10.1145/1458082.1458165 Cited on page(s) 68

P. Donmez, J. Carbonell, and J. Schneider. Efficiently learning the accuracy of labeling sources for selective sampling. In *Proceedings of the International Conference on Knowledge Discovery and Data Mining (KDD)*, pages 259–268. ACM, 2009. DOI: 10.1145/1557019.1557053 Cited on page(s) 74

P. Donmez, J. Carbonell, and J. Schneider. A probabilistic framework to learn from multiple annotators with time-varying accuracy. In *Proceedings of the SIAM Conference on Data Mining (SDM)*,
2010. Cited on page(s) 74

G. Druck, G. Mann, and A. McCallum. Learning from labeled features using generalized expectation
criteria. In *Proceedings of the ACM SIGIR Conference on Research and Development in Information
Retrieval*, pages 595–602. ACM, 2008. DOI: 10.1145/1390334.1390436 Cited on page(s) 69

G. Druck, B. Settles, and A. McCallum. Active learning by labeling features. In *Proceedings of the
Conference on Empirical Methods in Natural Language Processing (EMNLP)*, pages 81–90. ACL,
2009. DOI: 10.3115/1699510.1699522 Cited on page(s) 69

R. Duda, P. Hart, and D. Stork. *Pattern Classification*. Wiley-Interscience, 2001. Cited on page(s)
xi

V. Federov. *Theory of Optimal Experiments*. Academic Press, 1972. Cited on page(s) 18, 41

P. Felt, E. Ringger, K. Seppi, K. Heal, R. Haertel, and D. Lonsdale. First results in a study evaluating
pre-annotation and correction propagation for machine-assisted syriac morphological analysis.
In *Proceedings of the International Conference on Language Resources and Evaluation (LREC)*, pages
878–885. ELRA, 2012. Cited on page(s) 66

P. Flaherty, M. Jordan, and A. Arkin. Robust design of biological experiments. In *Advances in Neural
Information Processing Systems (NIPS)*, volume 18, pages 363–370. MIT Press, 2006. Cited on
page(s) 42

Y. Freund and R.E. Schapire. A decision-theoretic generalization of on-line learning and an application to boosting. *Journal of Computer and System Sciences*, 55(1):119–139, 1997.
DOI: 10.1006/jcss.1997.1504 Cited on page(s) 28

Y. Freund and R.E. Schapire. Large margin classification using the perceptron algorithm. *Machine
learning*, 37(3):277–296, 1999. DOI: 10.1023/A:1007662407062 Cited on page(s) 24

Y. Freund, H.S. Seung, E. Shamir, and N. Tishby. Selective samping using the query by committee
algorithm. *Machine Learning*, 28:133–168, 1997. DOI: 10.1023/A:1007330508534 Cited on
page(s) 28, 61, 62

E. Friedman. Active learning for smooth problems. In *Proceedings of the Conference on Learning
Theory (COLT)*, pages 3–2, 2009. Cited on page(s) 61

A. Fujii, T. Tokunaga, K. Inui, and H. Tanaka. Selective sampling for example-based word sense
disambiguation. *Computational Linguistics*, 24(4):573–597, 1998. Cited on page(s) 8, 49

K. Ganchev, J. Graça, J. Gillenwater, and B. Taskar. Posterior regularization for structured latent
variable models. *Journal of Machine Learning Research*, 11:2001–2049, 2010. Cited on page(s) 68

S. Geman, E. Bienenstock, and R. Doursat. Neural networks and the bias/variance dilemma. *Neural Computation*, 4:1–58, 1992. DOI: 10.1162/neco.1992.4.1.1 Cited on page(s) 40

Y. Grandvalet and Y. Bengio. Semi-supervised learning by entropy minimization. In *Advances in Neural Information Processing Systems (NIPS)*, volume 17, pages 529–536. MIT Press, 2005. Cited on page(s) 53

C. Guestrin, A. Krause, and A.P. Singh. Near-optimal sensor placements in gaussian processes. In *Proceedings of the International Conference on Machine Learning (ICML)*, pages 265–272. ACM, 2005. DOI: 10.1145/1102351.1102385 Cited on page(s) 45

Y. Guo and R. Greiner. Optimistic active learning using mutual information. In *Proceedings of International Joint Conference on Artificial Intelligence (IJCAI)*, pages 823–829. AAAI Press, 2007. Cited on page(s) 39

Y. Guo and D. Schuurmans. Discriminative batch mode active learning. In *Advances in Neural Information Processing Systems (NIPS)*, number 20, pages 593–600. MIT Press, Cambridge, MA, 2008. Cited on page(s) 45

R. Haertel, K. Seppi, E. Ringger, and J. Carroll. Return on investment for active learning. In *Proceedings of the NIPS Workshop on Cost-Sensitive Learning*, 2008. Cited on page(s) 67, 68

S. Hanneke. A bound on the label complexity of agnostic active learning. In *Proceedings of the International Conference on Machine Learning (ICML)*, pages 353–360. ACM, 2007. DOI: 10.1145/1273496.1273541 Cited on page(s) 59, 61, 62

S. Hanneke. *Theoretical Foundations of Active Learning*. PhD thesis, Carnegie Mellon University, 2009. Cited on page(s) 59, 62

A. Hauptmann, W. Lin, R. Yan, J. Yang, and M.Y. Chen. Extreme video retrieval: joint maximization of human and computer performance. In *Proceedings of the ACM Workshop on Multimedia Image Retrieval*, pages 385–394. ACM, 2006. DOI: 10.1145/1180639.1180721 Cited on page(s) 9

D. Haussler. Learning conjunctive concepts in structural domains. *Machine Learning*, 4(1):7–40, 1994. DOI: 10.1023/A:1022698821832 Cited on page(s) 26

S.C.H. Hoi, R. Jin, and M.R. Lyu. Large-scale text categorization by batch mode active learning. In *Proceedings of the International Conference on the World Wide Web*, pages 633–642. ACM, 2006a. DOI: 10.1145/1135777.1135870 Cited on page(s) 9, 43, 44

S.C.H. Hoi, R. Jin, J. Zhu, and M.R. Lyu. Batch mode active learning and its application to medical image classification. In *Proceedings of the International Conference on Machine Learning (ICML)*, pages 417–424. ACM, 2006b. DOI: 10.1145/1143844.1143897 Cited on page(s) 45

R. Hwa. On minimizing training corpus for parser acquisition. In *Proceedings of the Conference on Natural Language Learning (CoNLL)*, pages 1–6. ACL, 2001. DOI: 10.3115/1117822.1117829 Cited on page(s) 76

R. Hwa. Sample selection for statistical parsing. *Computational Linguistics*, 30(3):73–77, 2004. DOI: 10.1162/0891201041850894 Cited on page(s) 17

P. Jain, S. Vijayanarasimhan, and K. Grauman. Hashing hyperplane queries to near points with applications to large-scale active learning. In J. Lafferty, C. K. I. Williams, J. Shawe-Taylor, R.S. Zemel, and A. Culotta, editors, *Advances in Neural Information Processing Systems (NIPS)*, volume 23, pages 928–936. 2010. Cited on page(s) 24

J. Kang, K. Ryu, and H.C. Kwon. Using cluster-based sampling to select initial training set for active learning in text classification. In *Proceedings of the Pacific-Asia Conference on Advances in Knowledge Discovery and Data Mining (PAKDD)*, pages 384–388. Springer, 2004. DOI: 10.1007/978-3-540-24775-3_46 Cited on page(s) 49

A. Kapoor, E. Horvitz, and S. Basu. Selective supervision: Guiding supervised learning with decision-theoretic active learning,. In *Proceedings of International Joint Conference on Artificial Intelligence (IJCAI)*, pages 877–882. AAAI Press, 2007. Cited on page(s) 66

R.D. King, K.E. Whelan, F.M. Jones, P.G. Reiser, C.H. Bryant, S.H. Muggleton, D.B. Kell, and S.G. Oliver. Functional genomic hypothesis generation and experimentation by a robot scientist. *Nature*, 427(6971):247–52, 2004. DOI: 10.1038/nature02236 Cited on page(s) 7, 32, 66

R.D. King, J. Rowland, S.G. Oliver, M. Young, W. Aubrey, E. Byrne, M. Liakata, M. Markham, P. Pir, L.N. Soldatova, A. Sparkes, K.E. Whelan, and A. Clare. The automation of science. *Science*, 324(5923):85–89, 2009. DOI: 10.1126/science.1165620 Cited on page(s) 7

D. Koller and N. Friedman. *Probabilistic Graphical Models: Principles and Techniques*. MIT Press, 2009. Cited on page(s) 12, 17

C. Körner and S. Wrobel. Multi-class ensemble-based active learning. In *Proceedings of the European Conference on Machine Learning (ECML)*, pages 687–694. Springer, 2006. DOI: 10.1007/11871842_68 Cited on page(s) 63

A. Krause. *Optimizing Sensing: Theory and Applications*. PhD thesis, Carnegie Mellon University, 2008. Cited on page(s) 45

V. Krishnamurthy. Algorithms for optimal scheduling and management of hidden markov model sensors. *IEEE Transactions on Signal Processing*, 50(6):1382–1397, 2002. DOI: 10.1109/TSP.2002.1003062 Cited on page(s) 8

S. Kullback and R.A. Leibler. On information and sufficiency. *Annals of Mathematical Statistics*, 22: 79–86, 1951. DOI: 10.1214/aoms/1177729694 Cited on page(s) 29

G. Kunapuli, R. Maclin, and J. Shavlik. Advice refinement for knowledge-based support vector machines. In *Advances in Neural Information Processing Systems (NIPS)*, volume 24, pages 1728–1736. 2011. Cited on page(s) 68

J. Lafferty, A. McCallum, and F. Pereira. Conditional random fields: Probabilistic models for segmenting and labeling sequence data. In *Proceedings of the International Conference on Machine Learning (ICML)*, pages 282–289. Morgan Kaufmann, 2001. DOI: 10.1038/nprot.2006.61 Cited on page(s) 39

K. Lang. Newsweeder: Learning to filter netnews. In *Proceedings of the International Conference on Machine Learning (ICML)*, pages 331–339. Morgan Kaufmann, 1995. Cited on page(s) 49

K. Lang and E. Baum. Query learning can work poorly when a human oracle is used. In *Proceedings of the IEEE International Joint Conference on Neural Networks*, pages 335–340. IEEE Press, 1992. Cited on page(s) 6, 7

D. Lewis and J. Catlett. Heterogeneous uncertainty sampling for supervised learning. In *Proceedings of the International Conference on Machine Learning (ICML)*, pages 148–156. Morgan Kaufmann, 1994. Cited on page(s) 11, 76

D. Lewis and W. Gale. A sequential algorithm for training text classifiers. In *Proceedings of the ACM SIGIR Conference on Research and Development in Information Retrieval*, pages 3–12. ACM/Springer, 1994. DOI: 10.1145/62437.62470 Cited on page(s) 9

P. Liang, M.I. Jordan, and D. Klein. Learning from measurements in exponential families. In *Proceedings of the International Conference on Machine Learning (ICML)*, pages 641–648. Omnipress, 2009. DOI: 10.1145/1553374.1553457 Cited on page(s) 69

R. Liere and P. Tadepalli. Active learning with committees for text categorization. In *Proceedings of the Conference on Artificial Intelligence (AAAI)*, pages 591—597. AAAI Press, 1997. Cited on page(s) 28

Y. Liu. Active learning with support vector machine applied to gene expression data for cancer classification. *Journal of Chemical Information and Computer Sciences*, 44:1936–1941, 2004. DOI: 10.1021/ci049810a Cited on page(s) 9

Z. Lu and J. Bongard. Exploiting multiple classifier types with active learning. In *Proceedings of the Conference on Genetic and Evolutionary Computation (GECCO)*, pages 1905–1906. ACM, 2009. DOI: 10.1145/1569901.1570228 Cited on page(s) 77

D. MacKay. Information-based objective functions for active data selection. *Neural Computation*, 4 (4):590–604, 1992. DOI: 10.1162/neco.1992.4.4.590 Cited on page(s) 43, 45

G. Mann and A. McCallum. Efficient computation of entropy gradient for semi-supervised conditional random fields. In *Proceedings of the North American Association for Computational Linguistics (NAACL)*, pages 109–112. ACL, 2007. Cited on page(s) 17

G. Mann and A. McCallum. Generalized expectation criteria for semi-supervised learning of conditional random fields. In *Proceedings of the Association for Computational Linguistics (ACL)*, pages 870–878. ACL, 2008. Cited on page(s) 69

G.S. Mann and A. McCallum. Generalized expectation criteria for semi-supervised learning with weakly labeled data. *Journal of Machine Learning Research*, 11:955–984, 2010. Cited on page(s) 68

C. D. Manning and H. Schütze. *Foundations of Statistical Natural Language Processing*. MIT Press, 1999. Cited on page(s) 31, 52

O. Maron and T. Lozano-Perez. A framework for multiple-instance learning. In *Advances in Neural Information Processing Systems (NIPS)*, volume 10, pages 570–576. MIT Press, 1998. Cited on page(s) 71

A. McCallum and K. Nigam. Employing EM in pool-based active learning for text classification. In *Proceedings of the International Conference on Machine Learning (ICML)*, pages 359–367. Morgan Kaufmann, 1998. Cited on page(s) 9, 28, 49, 54

P. Melville and R. Mooney. Diverse ensembles for active learning. In *Proceedings of the International Conference on Machine Learning (ICML)*, pages 584–591. Morgan Kaufmann, 2004. DOI: 10.1145/1015330.1015385 Cited on page(s) 29

P. Melville, S.M. Yang, M. Saar-Tsechansky, and R. Mooney. Active learning for probability estimation using Jensen-Shannon divergence. In *Proceedings of the European Conference on Machine Learning (ECML)*, pages 268–279. Springer, 2005. DOI: 10.1007/11564096_28 Cited on page(s) 31

M. Mintz, S. Bills, R. Snow, and D. Jurafsky. Distant supervision for relation extraction without labeled data. In *Proceedings of the Association for Computational Linguistics (ACL)*, pages 1003–1011. ACL, 2009. Cited on page(s) 74

T. Mitchell. Generalization as search. *Artificial Intelligence*, 18:203–226, 1982. DOI: 10.1016/0004-3702(82)90040-6 Cited on page(s) 8, 21

T. Mitchell. *Machine Learning*. McGraw-Hill, 1997. Cited on page(s) xi, 57

R. Moskovitch, N. Nissim, D. Stopel, C. Feher, R. Englert, and Y. Elovici. Improving the detection of unknown computer worms activity using active learning. In *Proceedings of the German Conference on AI*, pages 489–493. Springer, 2007. DOI: 10.1007/978-3-540-74565-5_47 Cited on page(s) 9

I. Muslea, S. Minton, and C.A. Knoblock. Selective sampling with redundant views. In *Proceedings of the National Conference on Artificial Intelligence (AAAI)*, pages 621–626. AAAI Press, 2000. Cited on page(s) 29

I. Muslea, S. Minton, and C.A. Knoblock. Active + semi-supervised learning = robust multi-view learning. In *Proceedings of the International Conference on Machine Learning (ICML)*, pages 435–442. Morgan Kaufmann, 2002. Cited on page(s) 54

G.L. Nemhauser, L.A. Wolsey, and M.L. Fisher. An analysis of approximations for maximizing submodular set functions. *Mathematical Programming*, 14(1):265–294, 1978. DOI: 10.1007/BF01588971 Cited on page(s) 45

G. Ngai and D. Yarowsky. Rule writing or annotation: Cost-efficient resource usage for base noun phrase chunking. In *Proceedings of the Association for Computational Linguistics (ACL)*, pages 117–125. ACL, 2000. DOI: 10.3115/1075218.1075234 Cited on page(s) 31

H.T. Nguyen and A. Smeulders. Active learning using pre-clustering. In *Proceedings of the International Conference on Machine Learning (ICML)*, pages 79–86. ACM, 2004. DOI: 10.1145/1015330.1015349 Cited on page(s) 49

F. Olsson and K. Tomanek. An intrinsic stopping criterion for committee-based active learning. In *Proceedings of the Conference on Computational Natural Language Learning (CoNLL)*, pages 138–146. ACL, 2009. DOI: 10.3115/1596374.1596398 Cited on page(s) 77

G. Paaß and J. Kindermann. Bayesian query construction for neural network models. In *Advances in Neural Information Processing Systems (NIPS)*, volume 7, pages 443–450. MIT Press, 1995. Cited on page(s) 44

B. Pang, L. Lee, and S. Vaithyanathan. Thumbs up: Sentiment classification using machine learning techniques. In *Proceedings of the Conference on Empirical Methods in Natural Language Processing (EMNLP)*, pages 79–86. ACL, 2002. DOI: 10.3115/1118693.1118704 Cited on page(s) 69

G.J. Qi, X.S. Hua, Y. Rui, J. Tang, and H.J. Zhang. Two-dimensional active learning for image classification. In *Proceedings of the Conference on Computer Vision and Pattern Recognition (CVPR)*, pages 1–8. IEEE Press, 2008. DOI: 10.1109/CVPR.2008.4587383 Cited on page(s) 76

H. Raghavan, O. Madani, and R. Jones. Active learning with feedback on both features and instances. *Journal of Machine Learning Research*, 7:1655–1686, 2006. Cited on page(s) 68

R. Rahmani and S.A. Goldman. MISSL: Multiple-instance semi-supervised learning. In *Proceedings of the International Conference on Machine Learning (ICML)*, pages 705–712. ACM, 2006. DOI: 10.1145/1143844.1143933 Cited on page(s) 71

S. Ray and M. Craven. Supervised versus multiple instance learning: An empirical comparison. In *Proceedings of the International Conference on Machine Learning (ICML)*, pages 697–704. ACM, 2005. DOI: 10.1145/1102351.1102439 Cited on page(s) 71

R. Reichart, K. Tomanek, U. Hahn, and A Rappoport. Multi-task active learning for linguistic annotations. In *Proceedings of the Association for Computational Linguistics (ACL)*, pages 861–869. ACL, 2008. Cited on page(s) 76

E. Ringger, M. Carmen, R. Haertel, K. Seppi, D. Lonsdale, P. McClanahan, J. Carroll, and N. Ellison. Assessing the costs of machine-assisted corpus annotation through a user study. In *Proceedings of the International Conference on Language Resources and Evaluation (LREC)*, pages 3318–3324. ELRA, 2008. Cited on page(s) 67

D. Roth and K. Small. Active learning for pipeline models. In *Proceedings of the National Conference on Artificial Intelligence (AAAI)*, pages 683–688. AAAI Press, 2008. Cited on page(s) 76

N. Roy and A. McCallum. Toward optimal active learning through sampling estimation of error reduction. In *Proceedings of the International Conference on Machine Learning (ICML)*, pages 441–448. Morgan Kaufmann, 2001. Cited on page(s) 38, 39, 49

S. Russell and P. Norvig. *Artificial Intelligence: A Modern Approach*. Prentice Hall, second edition, 2003. Cited on page(s) xi

A.I. Schein and L.H. Ungar. Active learning for logistic regression: An evaluation. *Machine Learning*, 68(3):235–265, 2007. DOI: 10.1007/s10994-007-5019-5 Cited on page(s) 43, 44, 63

M.J. Schervish. *Theory of Statistics*. Springer, 1995. DOI: 10.1007/978-1-4612-4250-5 Cited on page(s) 42

G. Schohn and D. Cohn. Less is more: Active learning with support vector machines. In *Proceedings of the International Conference on Machine Learning (ICML)*, pages 839–846. Morgan Kaufmann, 2000. Cited on page(s) 24

H. Schütze, E. Velipasaoglu, and J.O. Pedersen. Performance thresholding in practical text classification. In *Proceedings of the Conference on Information and Knowledge Management (CIKM)*, pages 662–671. ACM, 2006. DOI: 10.1145/1183614.1183709 Cited on page(s) 20

R. Schwartz and Y.-L. Chow. The N-best algorithm: an efficient and exact procedure for finding the N most likely sentence hypotheses. In *Proceedings of the International Conference on Acoustics, Speech, and Signal Processing (ICASSP)*, pages 81–83. IEEE Press, 1990. DOI: 10.1109/ICASSP.1990.115542 Cited on page(s) 17

B. Settles. Active learning literature survey. Computer Sciences Technical Report 1648, University of Wisconsin–Madison, 2009. Cited on page(s) xi, xiii

B. Settles. Closing the loop: Fast, interactive semi-supervised annotation with queries on features and instances. In *Proceedings of the Conference on Empirical Methods in Natural Language Processing (EMNLP)*, pages 1467–1478. ACL, 2011. Cited on page(s) 69, 70

B. Settles and M. Craven. An analysis of active learning strategies for sequence labeling tasks. In *Proceedings of the Conference on Empirical Methods in Natural Language Processing (EMNLP)*, pages 1069–1078. ACL, 2008. DOI: 10.3115/1613715.1613855 Cited on page(s) 9, 17, 43, 44, 49, 63

B. Settles, M. Craven, and L. Friedland. Active learning with real annotation costs. In *Proceedings of the NIPS Workshop on Cost-Sensitive Learning*, 2008a. Cited on page(s) 4, 67, 68

B. Settles, M. Craven, and S. Ray. Multiple-instance active learning. In *Advances in Neural Information Processing Systems (NIPS)*, volume 20, pages 1289–1296. MIT Press, 2008b. Cited on page(s) 71, 72

H.S. Seung, M. Opper, and H. Sompolinsky. Query by committee. In *Proceedings of the ACM Workshop on Computational Learning Theory*, pages 287–294. ACM, 1992. DOI: 10.1145/130385.130417 Cited on page(s) 28, 62

C.E. Shannon. A mathematical theory of communication. *Bell System Technical Journal*, 27:379–423,623–656, 1948. Cited on page(s) 14

V.S. Sheng, F. Provost, and P.G. Ipeirotis. Get another label? improving data quality and data mining using multiple, noisy labelers. In *Proceedings of the International Conference on Knowledge Discovery and Data Mining (KDD)*, pages 614–622. ACM, 2008. DOI: 10.1145/1401890.1401965 Cited on page(s) 74

V. Sindhwani, P. Melville, and R.D. Lawrence. Uncertainty sampling and transductive experimental design for active dual supervision. In *Proceedings of the International Conference on Machine Learning (ICML)*, pages 953–960. Omnipress, 2009. DOI: 10.1145/1553374.1553496 Cited on page(s) 69

K. Small, B.C. Wallace, C.E. Brodley, and T.A. Trikalinos. The constrained weight space SVM: Learning with ranked features. In *Proceedings of the International Conference on Machine Learning (ICML)*, pages 865–872. Omnipress, 2011. Cited on page(s) 68, 69

B.C. Smith, B. Settles, W.C. Hallows, M.W. Craven, and J.M. Denu. SIRT3 substrate specificity determined by peptide arrays and machine learning. *ACS Chemical Biology*, 6(2):146–157, 2011. DOI: 10.1021/cb100218d Cited on page(s) 4

R. Snow, B. O'Connor, D. Jurafsky, and A. Ng. Cheap and fast—but is it good? In *Proceedings of the Conference on Empirical Methods in Natural Language Processing (EMNLP)*, pages 254–263. ACM, 2008. DOI: 10.3115/1613715.1613751 Cited on page(s) 74

M. Sugiyama and N. Rubens. Active learning with model selection in linear regression. In *Proceedings of the SIAM International Conference on Data Mining*, pages 518–529. SIAM, 2008. Cited on page(s) 77

C.A. Thompson, M.E. Califf, and R.J. Mooney. Active learning for natural language parsing and information extraction. In *Proceedings of the International Conference on Machine Learning (ICML)*, pages 406–414. Morgan Kaufmann, 1999. Cited on page(s) 9

K. Tomanek and U. Hahn. Semi-supervised active learning for sequence labeling. In *Proceedings of the Association for Computational Linguistics (ACL)*, pages 1039–1047. ACL, 2009. Cited on page(s) 54

K. Tomanek and U. Hahn. A comparison of models for cost-sensitive active learning. In *Proceedings of the International Conference on Computational Linguistics (COLING)*, volume Posters, pages 1247–1255. ACL, 2010. Cited on page(s) 68

K. Tomanek and K. Morik. Inspecting sample reusability for active learning. In *Active Learning and Experimental Design*, volume 15 of *JMLR Workshop and Conference Proceedings*, pages 169–181. Microtome Publishing, 2011. Cited on page(s) 76

K. Tomanek and F. Olsson. A web survey on the use of active learning to support annotation of text data. In *Proceedings of the NAACL HLT Workshop on Active Learning for Natural Language Processing*, pages 45–48. ACL, 2009. DOI: 10.3115/1564131.1564140 Cited on page(s) 64

K. Tomanek, J. Wermter, and U. Hahn. An approach to text corpus construction which cuts annotation costs and maintains reusability of annotated data. In *Proceedings of the Conference on Empirical Methods in Natural Language Processing (EMNLP)*, pages 486–495. ACL, 2007. Cited on page(s) 76

K. Tomanek, F. Laws, U. Hahn, and H. Schütze. On proper unit selection in active learning: Co-selection effects for named entity recognition. In *Proceedings of the NAACL HLT Workshop on Active Learning for Natural Language Processing*, pages 9–17. ACL, 2009. DOI: 10.3115/1564131.1564135 Cited on page(s) 20

S. Tong and E. Chang. Support vector machine active learning for image retrieval. In *Proceedings of the ACM International Conference on Multimedia*, pages 107–118. ACM, 2001. DOI: 10.1145/500141.500159 Cited on page(s) 9

S. Tong and D. Koller. Support vector machine active learning with applications to text classification. In *Proceedings of the International Conference on Machine Learning (ICML)*, pages 999–1006. Morgan Kaufmann, 2000. DOI: 10.1162/153244302760185243 Cited on page(s) 9, 23, 24

G. Towell and J. Shavlik. Knowledge-based artificial neural networks. *Artificial Intelligence*, 70: 119–165, 1994. DOI: 10.1016/0004-3702(94)90105-8 Cited on page(s) 68

G. Tür, D. Hakkani-Tür, and R.E. Schapire. Combining active and semi-supervised learning for spoken language understanding. *Speech Communication*, 45(2):171–186, 2005. DOI: 10.1016/j.specom.2004.08.002 Cited on page(s) 9, 54

A. Tversky and D. Kahneman. Extension versus intuitive reasoning: The conjunction fallacy in probability judgment. *Psychological Review*, 90(4):293—315, 1983. DOI: 10.1037/0033-295X.90.4.293 Cited on page(s) 69

L.G. Valiant. A theory of the learnable. *Communications of the ACM*, 27(11):1134–1142, 1984. DOI: 10.1145/1968.1972 Cited on page(s) 2, 57

V. Vapnik. *Statistical Learning Theory*. Wiley, 1998. Cited on page(s) 23, 57, 58

V.N. Vapnik and A. Chervonenkis. On the uniform convergence of relative frequencies of events to their probabilities. *Theory of Probability and Its Applications*, 16:264–280, 1971. DOI: 10.1137/1116025 Cited on page(s) 58

S. Vijayanarasimhan and K. Grauman. What's it going to cost you? Predicting effort vs. informativeness for multi-label image annotations. In *Proceedings of the Conference on Computer Vision and Pattern Recognition (CVPR)*, pages 2262–2269. IEEE Press, 2009a. Cited on page(s) 67, 68, 72

S. Vijayanarasimhan and K. Grauman. Multi-level active prediction of useful image annotations for recognition. In *Advances in Neural Information Processing Systems (NIPS)*, volume 21, pages 1705–1712. MIT Press, 2009b. Cited on page(s) 72

A. Vlachos. A stopping criterion for active learning. *Computer Speech and Language*, 22(3):295–312, 2008. DOI: 10.1016/j.csl.2007.12.001 Cited on page(s) 77

B.C. Wallace, K. Small, C.E. Brodley, and T.A. Trikalinos. Active learning for biomedical citation screening. In *Proceedings of the International Conference on Knowledge Discovery and Data Mining (KDD)*, pages 173–182. ACM, 2010a. DOI: 10.1145/1835804.1835829 Cited on page(s) 20, 72

B.C. Wallace, K. Small, C.E. Brodley, and T.A. Trikalinos. Who should label what? Instance allocation in multiple expert active learning. In *Proceedings of the SIAM Conference on Data Mining (SDM)*, pages 176–187, 2011. Cited on page(s) 74

Byron C. Wallace, Kevin Small, Carla E. Brodley, Joseph Lau, and Thomas A. Trikalinos. Modeling annotation time to reduce workload in comparative effectiveness reviews. In *Proceedings of the ACM International Health Informatics Symposium (IHI)*, pages 28–35. ACM, 2010b. DOI: 10.1145/1882992.1882999 Cited on page(s) 67, 68

L. Wang. Sufficient conditions for agnostic active learnable. In *Advances in Neural Information Processing Systems (NIPS)*, volume 22, pages 1999–2007. 2009. Cited on page(s) 61

J.H. Ward. Hierarchical grouping to optimize an objective function. *Journal of the American Statistical Association*, 58:236–244, 1963. DOI: 10.2307/2282967 Cited on page(s) 52

M. Warmuth, K. Glocer, and G. Rätsch. Boosting algorithms for maximizing the soft margin. In *Advances in Neural Information Processing Systems (NIPS)*, volume 20, pages 1585–1592. MIT Press, 2008. Cited on page(s) 24

Z. Xu, R. Akella, and Y. Zhang. Incorporating diversity and density in active learning for relevance feedback. In *Proceedings of the European Conference on IR Research (ECIR)*, pages 246–257. Springer-Verlag, 2007. DOI: 10.1007/978-3-540-71496-5_24 Cited on page(s) 44, 49

R. Yan, J. Yang, and A. Hauptmann. Automatically labeling video data using multi-class active learning. In *Proceedings of the International Conference on Computer Vision*, pages 516–523. IEEE Press, 2003. DOI: 10.1109/ICCV.2003.1238391 Cited on page(s) 9

D. Yarowsky. Unsupervised word sense disambiguation rivaling supervised methods. In *Proceedings of the Association for Computational Linguistics (ACL)*, pages 189–196. ACL, 1995. Cited on page(s) 53

H. Yu. SVM selective sampling for ranking with application to data retrieval. In *Proceedings of the International Conference on Knowledge Discovery and Data Mining (KDD)*, pages 354–363. ACM, 2005. DOI: 10.1145/1081870.1081911 Cited on page(s) 8

K. Yu, J. Bi, and V. Tresp. Active learning via transductive experimental design. In *Proceedings of the International Conference on Machine Learning (ICML)*, pages 1081–1087. ACM, 2006. DOI: 10.1145/1143844.1143980 Cited on page(s) 54

C. Zhang and T. Chen. An active learning framework for content based information retrieval. *IEEE Transactions on Multimedia*, 4(2):260–268, 2002. DOI: 10.1109/TMM.2002.1017738 Cited on page(s) 9

T. Zhang and F.J. Oles. A probability analysis on the value of unlabeled data for classification problems. In *Proceedings of the International Conference on Machine Learning (ICML)*, pages 1191–1198. Morgan Kaufmann, 2000. Cited on page(s) 43

Y. Zhang. Multi-task active learning with output constraints. In *Proceedings of the Conference on Artificial Intelligence (AAAI)*, pages 667–672. AAAI Press, 2010. Cited on page(s) 75

Z.H. Zhou, K.J. Chen, and Y. Jiang. Exploiting unlabeled data in content-based image retrieval. In *Proceedings of the European Conference on Machine Learning (ECML)*, pages 425–435. Springer, 2004. Cited on page(s) 54

X. Zhu and A. Goldberg. *Introduction to Semi-Supervised Learning*. Synthesis Lectures on Artificial Intelligence and Machine Learning. Morgan & Claypool, 2009. Cited on page(s) 4, 53

X. Zhu, J. Lafferty, and Z. Ghahramani. Combining active learning and semi-supervised learning using Gaussian fields and harmonic functions. In *Proceedings of the ICML Workshop on the Continuum from Labeled to Unlabeled Data*, pages 58–65, 2003. Cited on page(s) 39, 54

Author's Biography

BURR SETTLES

Burr Settles is a Postdoctoral Research Scientist in the Machine Learning Department at Carnegie Mellon University. He received a PhD in computer sciences from the University of Wisconsin-Madison in 2008, with additional studies in linguistics and biology. His research focuses on machine learning technology that interacts with humans as part of knowledge acquisition and training, with applications in natural language processing, biological discovery, and social computing.

Burr is the author of more than 20 research papers, including a popular literature survey on active learning (which was the genesis for this book). He has organized workshops at the International Conference on Machine Learning (ICML) and the North American Chapter of the Association for Computational Linguistics (NAACL) on active learning and related topics in cost-sensitive machine learning. In his spare time, he runs a website community for musicians (FAWM.ORG), prefers sandals to shoes, and plays guitar in the Pittsburgh pop band Delicious Pastries.

Index

Printed in the United States
by Baker & Taylor Publisher Services